Molecular Neurotoxicology

Environmental Agents and Transcription-Transduction Coupling

Molecular Neurotoxicology

Environmental Agents and Transcription-Transduction Coupling

Edited by

Nasser H. Zawia, Ph.D.

University of Rhode Island
Kingston, RI

CRC PRESS

Boca Raton   London   New York   Washington, D.C.

Library of Congress Cataloging-in-Publication Data

Molecular toxicology : environmental agents and transcription-transduction coupling/
edited by Nasser H. Zawia.
    p. cm.
Includes bibliographical references.
ISBN 0-415-28031-1 (alk. paper)
  1. Neurotoxicology. 2. Molecular toxicology. I. Zawia, Nasser H.

RC347.5.M655 2004
616.8'0471—dc22

2003070025

Visit the CRC Press Web site at www.crcpress.com

Preface

The nervous system is an organ whose primary function is to process, store, and transmit information. Although the final output of brain function is behavior, such an output is a culmination of a series of biochemical and molecular events, including the expression of genes and their translation into products. These molecular events are all potential targets for modulation by environmental agents. Molecules involved in these events provide a highly sensitive system that allows neurotoxicologists to detect adverse effects at low exposure levels. The availability of whole genome sequences and the development of new tools to identify and monitor transcriptional activity have accelerated the rate of discovery of new targets and endpoints of neurotoxicity. The discovery of new genes and the ability to monitor the activity of multiple genes that act in concert have now become possible. These new approaches have resulted in a reevaluation of the perceived health risks posed by environmental agents and the levels of concern previously thought to be adequate. Understanding the relationship between toxicant exposure and perturbations in gene transcription provides an in-depth mechanistic examination of how neurotoxicants may result in long-term perturbations of the nervous system. This book presents a historical overview which summarizes the major developments in "molecular neurotoxicology" and discusses the basics of molecular biology, including a survey of the tools commonly used. The book also contains chapters that address the role of specific transcription factor families in neurotoxicity, various models of growth and neurodegeneration, the role of DNA repair in the nervous system, and the impact of neurotoxicants on a variety of cell signaling intermediates.

The Editor

Nasser H. Zawia, Ph.D., is an Associate Professor of Toxicology at the University of Rhode Island and Assistant Director of the Rhode Island Biomedical Research Infrastructure Network (RI-BRIN). He has previously served as an Assistant Professor at the University of Sana'a, Yemen and at Meharry Medical College in Nashville, Tennessee.

He obtained his Ph.D. in pharmacology and toxicology from the University of California, Irvine; his MS in pharmacology and physiology from Loma Linda University; and his BS in biochemistry from the University of Massachusetts, Amherst. He received postdoctoral training at the National Institutes of Health (NIH)-National Institute of Environmental Health Sciences (NIEHS).

Dr. Zawia pioneered the application of the tools of molecular biology to the study of environmental neurotoxicology. He has been responsible for the delineation of zinc finger protein transcription factors as mediators of metal-induced perturbations in gene transcription.

Dr. Zawia has received numerous national and international awards and medals for his teaching and research excellence. He has served on many national panels and has also served as a grant reviewer for NIH and reviewer for many scientific journals. He has chaired many sessions at national meetings and has helped to organize regional, national, and international conferences. Dr. Zawia is a member of the Society of Toxicology, the International Neurotoxicology Association, the American Association for the Advancement of Science, the Society for Neuroscience, and Sigma Xi Scientific Honor Society.

Contributors

Saleh A. Bakheet
Department of Biomedical Sciences
University of Rhode Island
Kingston, Rhode Island

Md. Riyaz Basha
Department of Biomedical Sciences
College of Pharmacy
University of Rhode Island
Kingston, Rhode Island

Stephen Bondy
Department of Community and
  Environmental Medicine
Center for Occupational and
  Environmental Health
University of California
Irvine, California

T. L. Butler
Department of Pharmacology
College of Medicine
University of South Florida
Tampa, Florida

Arezoo Campbell
Department of Community and
  Environmental Medicine
Center for Occupational and
  Environmental Health
University of California
Irvine, California

Fernando Cardozo-Pelaez
Center for Environmental Health
  Sciences
Department of Pharmaceutical Sciences
University of Montana
Missoula, Montana

Tomás R. Guilarte
Environmental Health Sciences
Johns Hopkins University
Bloomberg School of Public Health
Baltimore, Maryland

C.A. Kassed
Department of Pharmacology
College of Medicine
University of South Florida
Tampa, Florida

Prasada Rao S. Kodavanti
Cellular and Molecular Toxicology
  Branch
Neurotoxicolgy
  Division/NHEERL/ORD
U.S. Environmental Protection Agency
Research Triangle Park, North Carolina

Simon Long
KS Biomedix
Guildford, Surrey, England

Keith Pennypacker
Department of Pharmacology
College of Medicine
University of South Florida
Tampa, Florida

G. R. Reddy
Division of Neurobiology
Department of Zoology
S.V. University
Tirupati, India

Todd Stedeford
Department of Neurology
College of Medicine
University of South Florida
Tampa Florida
and
Polish Academy of Sciences
Gliwice, Poland

Mohan C. Vemuri
Department of Surgery
Children's Hospital of Philadelphia
Philadelphia, Pennsylvania

Wei Wei
Department of Biomedical Sciences
College of Pharmacy
University of Rhode Island
Kingston, Rhode Island

Rajgopal Yadavalli
Sanders Brown Center on Aging
University of Kentucky
Lexington, Kentucky

Binfang Yan
Department of Biomedical Sciences
University of Rhode Island
Kingston, Rhode Island

Contents

CHAPTER 1

Introduction to Molecular Neurotoxicology

Nasser H. Zawia

CONTENTS

1.1 INTRODUCTION

The nervous system is an organ whose primary function is to process, store, and transmit information. As the central computing facility of the body, the nervous system is organized so as to handle the enormous volumes of information it receives from numerous sites within the body. The complex and hierarchical cytoarchitecture of the brain facilitates vital communication among cells, which ultimately determines the behavioral response of an organism. Although the final output of brain function is behavior, such an output is a culmination of a series of biochemical and molecular events, including the expression of genes and their translation into products. These molecular events are all potential targets for modulation by environmental agents. While the number of target genes and products studied following neurotoxicant exposure is numerous, it is important to focus attention on the effects of neurotoxic agents on signaling and transcriptional regulation. Understanding the relationship between toxicant exposure, signal transduction/transcription coupling, and perturbations in gene transcription provides an in-depth mechanistic examination of how neurotoxicants may result in long-term perturbations of the nervous system.

1.2 GENE EXPRESSION IN THE BRAIN: THEN AND NOW

The informational platform of the nervous system resides mainly in the connections between individual neurons and the pathways they represent (circuitry). The assembly of these neural networks and their long-term sustenance are dependent, however, on data mining from another information storage site within the cell, namely, the genome. Neuroscientists have long recognized that maintenance of the highly specialized functions of brain cells, their unique metabolic needs, and their structural properties required a vast array of genes and their products. Furthermore, the intricate development and growth of the brain follows a program in which the networking and connectivity of cells in multiple locations in the body is accomplished through temporal and spatial scheduling to ensure that both targets and processes are coordinated in their growth.

While the requirements for building and maintaining neural networks were understood, the tools for assessing the molecular complexity of brain cells were missing. Studies in the late 1970s and early 1980s, using basic nucleic acid hybridization techniques, began the cumbersome job of determining the level of gene expression in the central nervous system. Kinetic nucleic acid hybridization experiments validated the above assumptions and revealed that the brain expressed two to three times more mRNAs than other organs such as the liver or kidney (Chikaraishi et al., 1978; Chikaraishi, 1979; Sutcliffe, 1988). It was further observed that most of these messages were large (≈ 5 kb), and of low abundance. These preliminary techniques estimated that the number of brain-specific messages expressed was about 30,000. However, only 39 brain-specific genes were identified (Sutcliffe, 1988).

Using oligo (dT) columns and the like, it was soon discovered that a large pool of these messages were lacking a polyadenylation signal ($\approx 50\%$). Since the percentage of poly(A-) messages in other organs does not normally exceed 5%, the role and purpose of these poly(A-) mRNAs remained enigmatic (Brilliant et al., 1984). The poly(A) tail, which is attached to most messages, is believed to play a role in the stability and turnover of mRNAs. Since poly(A-) mRNAs appeared mostly postnatally, it was suspected that the absence of a polyadenylation signal might be related to the short-lived nature of these mRNAs. It was thus assumed that these messages served in a temporary manner to regulate specific steps during development of the brain.

Today, the entire human genome has been sequenced (Venter et al., 2001). Remarkably, the numbers of genes encoded in the genome (20,000–40,000) are similar to earlier estimates done for the brain (Sutcliffe et al., 1984). Within this range are about 26,383 annotated genes and 12,731 hypothetical ones. The most common genes code for proteins that interact directly with the genome itself, such as transcription factors and DNA enzymes (Venter et al., 2001). However, about 60% of encoded genes are still of unknown function (Table 1.1).

The information present in the genome of every cell in the body is identical. The set of specific genes expressed in a cell determine its differentiated state and specialized function. The sequencing of the human genome has indeed confirmed that gene expression in the brain is complex. When examining gene expression in the brain using current methodology, one finds that there is a marked increase in

Table 1.1 **Functional Representation of the Human Genome**

Gene Family	Percent of Genome
Nucleic acid binding	13.5
Transcription factors (6%)	
DNA-enzymes (7.5%)	
Signal transduction	13.5
Signaling molecules (1.2%)	
Receptors (5%)	
Kinases (2.8%)	
Select regulatory molecules (3.2%)	
Ion channels (1.3%)	
Metabolic enzymes	10.2
Other	2.8
Unknown function	60

Compiled from Venter et al., 2001.

members of protein families. Many protein families particularly involved in neural development have more members expressed in the brain than in other tissues (Cravchik et al., 2001). For example, more neurotrophic factors and signaling molecules are expressed in the human brain than any other organ or species (semaphorins: 6 in flies, 22 in humans). The nervous system also expresses a vast array of gene products to maintain its structure and function, such as myelin proteins, ion channels, synaptic proteins, cell adhesion molecules (connexins), axon guidance (ephrins, semaphorins, neuropilins, plexins), and notch receptors. Many of these molecules are expectedly absent from the transcriptome of most cells.

1.3 REGULATION OF GENE EXPRESSION

The number of genes encoded in the human genome is substantial. However, only subsets of these genes are expressed at any given time in any particular cell. The regulation of the expression of these genes in a coordinated manner and in response to various stimuli requires elaborate and highly sophisticated machinery. Differential gene expression is the hallmark of development and the essence of a living dynamic cell. The main players of this gene regulatory system are positive-acting (stimulatory) sequence-specific DNA-binding proteins known as transcription factors (TFs). The importance of these factors and their role in the multiple tiered steps of transcription is underlined by their representation in the human genome ($\approx$ 2000 genes code for TFs).

The large number of transcription factors present in each cell presents a challenge in terms of understanding their role and the conditions necessary for their action. It is important to distinguish between two categories of TFs. The first category are general transcription factors (GTFs) that bind the TATA box and are part of the core promoter, where RNA polymerase binds to initiate transcription of genes. Members of the second category are TFs that bind specific *cis*-DNA elements and modulate the rate of transcription of the initiation complex. See Chapter 2 for more detail.

Table 1.2   Structural Classification of Transcription Factors

Gene Family	Percent of Genome
Zinc containing modules	
Classical zinc fingers	Sp1, Sp2
(Cys_2-His_2, zinc finger)	Ets
	Egr-1, Egr-2
Nuclear receptors superfamily	ER
(Cys_4, zinc finger)	PR
	GR
	RAR
	VDR
POU homeodomains	Oct-1, Oct-2
(Octamer motifs)	Pit-1
Leucine zippers motifs	CREB
(ZIP)	AP-1
Helix-loop-helix motifs	c-myc
(HLH)	MyoD

Note: Sp, specificity protein, Ets, erythroblastosis virus oncogene homology; Egr, early growth response gene; ER, estrogen receptor; GR, glucocorticoid receptor; PR, progesterone receptor; RAR, retinoic acid receptor; VDR, vitamin D3 receptor; Oct, octamer transcription factor; Pit-1, pituitary-specific transcription factor; CREB, cAMP responsive element-binding protein; AP-1, activator protein-1; MyoD, myoblast determination factor.

One approach to grouping such factors relied on identifying structural similarities among them. Typically these proteins have two characteristic domains: a sequence-specific DNA-binding domain and a transactivation domain. Therefore, TFs were classified according to common structural elements present mainly in their DNA-binding domains (Harrison, 1991; Pabo and Sauer, 1992). Such a grouping placed TFs in one of four families: zinc finger proteins (Cys_2His_2: Sp1; Cys_4: ER, GR), POU domain proteins (pit-1), leucine zipper proteins (AP-1), and helix-loop-helix proteins (MyoD) (Table 1.2). While such a classification defined the manner of interaction of a specific TF with its corresponding DNA consensus elements and helped group numerous identified factors into families, it has some serious short-fallings. Chief among these is that such an approach would place two different TFs, which might be connected to different signal transduction pathways, under the same family. Such a grouping leads to the mistaken assumption that structurally similar members of a TF family probably regulate the same events in the cell.

A newly proposed alternative grouping classifies TFs according to their connections with cellular regulatory circuits (Brivanlou and Darnell, 2002). This takes into account extracellular and intracellular signaling pathways and their relationship to each TF. This approach provides a functional perspective of each TF and yields a better understanding of the role of that TF. At the basic level, this classification distinguishes between TFs that are "constitutive" such as Sp1, and those whose activation is "conditional," i.e., dependent on some signaling event. The "conditional" TFs are either those transiently expressed during development (Hox, Pit-1, GATA), or in response to signaling molecules. The signal-dependent (conditional)

Table 1.3 Functional Classification of Transcription Factors

Pathway	Activated By	Nuclear Factor
Steroid receptors	Glucocorticoids, estrogen, etc.	GR, ER, PR
Developmental	Growth factors	Oct-1, Oct-2, Sp1?
Resident in the nucleus*		
p53	Internal signals (DNA damage)	p53
cell surface receptors	Neurotransmitters	CREB, AP-1, ATM
Resident in the cytoplasm**		
SMAD	TGF-beta (serine kinases)	FAST1
STAT/JAK	Cytokines, growth factors	JAK
Rel/NFκB	Extracellular proteins	NFκB
CI/GLI	Hh ligands	Ci
Wnt	Wnt/Wg ligands	LEF/CAT
Notch	Membrane associated ligands	NICD/HLH
NFAT	Calcium	NFAT/AP-1
PLC	PI turnover	Tu

* Constitutively present in the nucleus, however, their activity is regulated by external signals.
**These TFs are latently present in the cytoplasm and must translocate to the nucleus. HLH, helix-loop-helix; CAT, catenin; ATM, ataxia-telangiectasia mutant; FAST1, forkhead activin signal transducer-1; JAK, Janus kinases; NFκB, nuclear factor kappa B; Ci, cubitus interruptus; NICD, Notch-1 intracellular domain; NFAT, nuclear factors in activated T cells; Tu, Tubby. Modified from Brivanlou and Darnell, 2002.

TFs fall into four groups: (1) the steroid receptor superfamily (GR, ER, PR, RXRs, RARs, etc.); (2) development-specific factors (Oct-1, Oct-2); (3) those respondent to internal signals (p53); and (4) those linked to cell-surface receptors (CREB, AP-1, NFκB, STATs etc.).

The largest and most complex class of "conditional" TFs is linked to cell surface receptor-ligand-signaling pathways. About eight major signaling pathways deliver an active transcription factor to the nucleus (Table 1.3). While the activation of all these TFs are dependent on extracellular signals, it is important to distinguish between those resident in the nucleus (CREB, AP-1) and those latently present in the cytoplasm which must then translocate to the nucleus (NFκB, STATs). These TFs mediate the activation of gene expression in response to a vast number of polypeptide signals such as growth factors, cytokines, charged small molecules that trigger tyrosine kinases, membrane associated ligands, and ligands that are coupled to various second messenger pathways (Brivanlou and Darnell, 2002).

The above discussed TFs act positively to increase the rate of transcription of genes. However, there are also subsets of proteins that act negatively to inhibit transcription. While a small number of these are site-specific DNA-binding proteins, the majority in eukaryotic systems are not gene-specific. These proteins often interact or interfere with co-activators or positively acting transcription factors, or physically block areas of DNA where activators would bind (for a review see Smith and Johnson, 2000). Furthermore, positive-acting or stimulatory TFs discussed above can also perform inhibitory roles. One way TFs can accomplish this function is by not possessing an activation domain. When a stimulatory TF lacks a transactivation domain and binds to its consensus element on the DNA, it can act as a repressor of gene expression. Therefore some factors can play dual roles depending on their constitution and on the presence of requisite domains.

1.4 SIGNALING TARGETS IN NEUROTOXICITY

Prior to the development of molecular biology tools, the majority of studies in the early 1970s dealt with examining the action of neurotoxicants on biochemical events in the cells, such as enzyme function or the influence of neurotoxicants on cell-surface receptors or the binding of ligands to them. As the various intermediates in cell signaling became identified and the circuitry that links various cell signaling pathways to the genome was defined, it became more apparent that signal transduction was a major site of action for neurotoxicants (see Chapters 5 and 8).

1.5 TRANSCRIPTIONAL TARGETS IN NEUROTOXICITY

It has become more apparent that exposure to various chemicals and environmental hazards elicits changes in the expression of a variety of genes. The study of gene expression and transcriptional regulation is an important aspect of understanding the mechanisms associated with neurotoxicity. In the 1990s, molecular neurotoxicologists began studying the steady-state levels of a variety of brain-specific mRNAs in response to various chemical exposures. Some of these genes included myelin genes (MBP, PLP, CNP), glial fibrillary acidic protein (GFAP), growth associated protein 43 (GAP-43), N-methyl-D-aspartate receptors (NMDAR1), and choline acetyltransferase (Zawia and Harry, 1995, 1996; Harry et al., 1996; Schmitte et al., 1996; Sun et al., 1997; Guilarte and McGlothan, 1998; Zawia et al., 1998). Now a variety of approaches such as utilization of genetically altered animals and cells, and the more recent developments of neurotoxicant screening using DNA and protein microarray techniques, are being utilized (Bouton et al., 2001; Bakheet and Zawia, 2002). The availability of whole genome sequences and the development of new tools to identify and monitor transcriptional activity have accelerated the rate of discovery of new targets and endpoints of neurotoxicity. The identification of new genes and the monitoring of the activity of multiple genes that act in concert have now become possible.

One of the earliest identifications of the interactions between neurotoxic compounds and gene transcription involved the study of the role of the Sp family of transcription factors and their mediation of metal-induced changes of gene expression in the developing brain (Zawia et al., 1998). Sp1 is a transcription factor which contains a zinc finger motif and whose activity is modulated both *in vivo* and *in vitro* by heavy metals such as Pb, Hg, and Cd (Razmiafshari and Zawia, 2000; Crumpton et al., 2001). Interactions of heavy metals with this family of transcription factors will be presented in more detail in Chapter 3 as an example of a transcriptional target modulated by environmental agents.

The AP-1 transcription factor and components of AP-1 such as Fos were some of the earliest transcription factors discovered in neuroscience (Morgan and Curran,

1986). The discovery of AP-1 and its activity-dependent coupling of signal-transduction to gene transcription opened the first window towards understanding how long-term adaptation in the nervous system may be brought about (Curran and Morgan, 1987). The activation of AP-1 in neurons that survive injury following ischemia, electroconvulsive shock, or exposure to neurotoxic metals such as trimethyltin or Pb suggested the important role of this TF family in regulating gene expression following toxicant exposure (Zawia and Bondy, 1990; Zawia et al., 1990; Vendrell et al., 1991; Zawia and Harry, 1993; Pennypacker et al., 1997; Chakraborti et al., 1999). Another transcription factor involved in neuronal survival and repair following neurotoxic injury is NFκB (Pyatt et al., 1996; Ramesh et al., 1999; Pennypacker et al., 2000). While AP-1 is resident in the nucleus, NFκB must translocate to the nucleus to turn on gene expression. Therefore neurotoxicants target different signaling intermediates to influence gene expression regulated by these factors.

In addition to the role of TFs in adjusting gene expression associated with neurons, TFs are also active in regulating gene expression in glial cells. Glial activation, or reactive gliosis, represents a sensitive and early response of the nervous system to all types of neurotoxic injuries (O'Callaghan, 1998). While all the above TFs may be present in glia and may be activated following neurotoxic injury, an important pathway linking signal transduction to changes in gene expression following neurotoxic insults is the STAT/JAK pathway (Little et al., 2002). This pathway, which normally mediates the actions of a variety of cytokines and growth factors, is also subject to modulation by neurotoxicants (Hebert and O'Callaghan, 2000).

While the study of perturbations of TFs is very useful from a mechanistic point of view, it is still important to identify the genes that may be targeted by these TFs following a neurotoxic exposure. While this process is cumbersome, it has been greatly simplified by the use of DNA-array technology and the sequencing of entire genomes of various organisms. One can now screen a large number of genes and determine the elevated expression of some genes following an exposure scenario. Once a target gene has been identified, the next step is to examine the regulatory regions of that gene. If it is a known gene that has been sequenced, a glance at its promoter region will reveal the presence of consensus elements recognized by various TFs. These promoter regions often contain multiple consensus elements for numerous TFs. The presence of a response element for TF is an initial evidence for the ability of that TF to drive the expression of that gene. However, the presence of a response element in itself does not imply that the corresponding TF is responsible for the activation of that gene. Many response elements on genes are silent and very often the activation of a gene requires the combinatorial action of multiple TFs.

1.6 CONCLUSIONS

Gene transcription mediates the long-term adaptation of cells to aversive stimuli. Therefore, examining the involvement of transcriptional events as an endpoint of neurotoxicity is crucial to understanding lasting effects that may be a product of exposure to neurotoxic agents. It is also essential to know the unique aspects of transcription in the brain and the specific steps that may be targeted by a neurotoxicant in signal transduction/transcription coupling following an exposure. Such an in-depth and molecular examination will bring about a better understanding of the mechanisms involved in neurotoxicity.

REFERENCES

Bakheet, S. and Zawia, N.H., Temporal changes in the expression of transcription factors in the developing lead-exposed hippocampus as determined by macroarray analysis, *Toxicologist*, 66, 1647, 2002.

Brilliant, M.H., Sueoka, N., and Chikaraishi, D.M., Cloning of DNA corresponding to rare transcripts of rat brain: evidence of transcriptional and post-transcriptional control and of the existence of nonpolyadenylated transcripts, *Mol. Cell. Biol.*, 4, 2187–2197, 1984.

Bouton, C.M. et al., Microarray analysis of differential gene expression in lead-exposed astrocytes, *Toxicol. Appl. Pharmacol.*, 176, 34–53, 2001.

Brivanlou, A.H., and Darnell, J.E., Signal transduction and the control of gene expression, *Science*, 295, 813–818, 2002.

Chakraborti, T. et al., Increased AP-1 DNA binding activity in PC12 cells treated with lead, *J. Neurochem.*, 73, 187–194, 1999.

Chikaraishi, D.M., Deeb, S.S., and Sueoka, N., Sequence complexity of nuclear RNAs in adult rat tissues, *Cell*, 13, 111–120, 1978.

Chikaraishi, D.M., Complexity of cytoplasmic polyadenylated and nonpolyadenylated rat brain ribonucleic acids, *Biochemistry*, 24, 3249–3256, 1979.

Cravchik, A. et al., Sequence analysis of the human genome: implications for the understanding of nervous system function and disease, *Arch. Neurol.*, 58, 1772–1778, 2001.

Crumpton, T. et al., Lead exposure in pheochromocytoma (PC12) cells alters neural differentiation and Sp1 DNA-binding, *NeuroToxicology*, 22, 49–62, 2001.

Curran, T., and Morgan, J.I., Memories of fos, *Bioessays*, 7, 255–258, 1987.

Guilarte, T.R., and McGlothan, J.L., Hippocampal NMDA receptor mRNA undergoes subunit specific changes during developmental lead exposure, *Brain Res.*, 790, 98–107, 1998.

Harrison, S.C., A structural taxonomy of DNA-binding domains, *Nature*, 353, 715–719, 1991.

Harry, G.J. et al., Pb-induced alterations of glial fibrillary acidic protein (GFAP) in the developing rat brain, *Toxicol. Appl. Pharmacol.*, 13, 9, 84–93, 1996.

Hebert, M.A., and O'Callaghan, J.P., Protein phosphorylation cascades associated with methamphetamine-induced glial activation, *Ann. N. Y. Acad. Sci.*, 914, 238–262, 2000.

Little, A.R. et al., Chemically induced neuronal damage and gliosis: enhanced expression of the proinflammatory chemokine, monocyte chemoattractant protein (MCP)-1, without a corresponding increase in proinflammatory cytokines(1), *Neuroscience*, 115, 307–320, 2002.

Morgan, J.I., and Curran, T., Role of ion flux in the control of c-fos expression, *Nature*, 322, 552–555, 1986.

O'Callaghan, J.P., Astrocytes: key players in mediation or modulation of neurotoxic responses? Commentary on forum position paper, *NeuroToxicology*, 19, 35–36, 1998.

Pabo, C.O., and Sauer, R.T., Transcription factors: structural families and principles of DNA recognition, *Annu. Rev. Biochem.*, 61, 1053–1095, 1992.

Pennypacker, K.R. et al., Lead-induced developmental changes in AP-1 DNA binding in rat brain, *Int. J. Dev. Neurosci.*, 15, 321–328, 1997.

Pennypacker, K.R. et al., Brain injury: prolonged induction of transcription factors, *Acta Neurobiol. Exp. (Warsz)*, 60, 515–530, 2000.

Pyatt, D.W. et al., Inorganic lead activates NF-kappa B in primary human CD4+ T lymphocytes, *Biochem. Biophys. Res. Commun.*, 227, 380–385, 1996.

Ramesh, G., et al., Lead activates nuclear transcription factor-kappa beta, activator protein-1 and amino-terminal c-Jun kinase in pheochromocytoma cells, *Toxicol. Appl. Pharmacol.*, 155, 280–286, 1999.

Razmiafshari, M., and Zawia, N.H., Utilization of a synthetic peptide as a tool to study the interaction of heavy metals with the zinc finger domain of proteins critical for gene expression in the developing brain, *Toxicol. Appl. Pharmacol.*, 166, 1–12, 2000.

Schmitte, T., Zawia, N.H., and Harry, G.J., GAP-43 mRNA expression in the developing rat brain: alterations following lead-acetate exposure, *NeuroToxicology*, 17, 407–414, 1996.

Smith, R.L., and Johnson, A.D., Turning genes off by Ssn6-Tup1: a conserved system of transcriptional repression in eukaryotes, *Trends Biochem. Sci.*, 25, 325–330, 2000.

Sutcliffe, J.G. et al., Control of neuronal gene expression, *Science*, 225, 1308–1315, 1984.

Sutcliffe, J.G., mRNA in the mammalian central nervous system, *Annu. Rev. Neurosci.*, 11, 157–198, 1988.

Sun, X., Tian, X., and Suszkiw, J.B., Reduction of choline acetyltransferase mRNA in the septum of developing rat exposed to inorganic lead, *NeuroToxicology*, 18, 201–207, 1997.

Vendrell, M. et al., *c-fos* and ODC gene expression in brain as early markers of neurotoxicity, *Brain Res.*, 544, 291–296, 1991.

Venter, J.C. et al., The sequence of the human genome, *Science*, 291, 1304–1351, 2001.

Zawia, N.H., and Bondy, S.C., Activity-dependent rapid gene expression in the brain: ornithine decarboxylase and *c-fos*, *Mol. Brain Res.*, 7, 243–247, 1990.

Zawia, N.H., Vendrell, M. and Bondy, S.C., Potentiation of stimulated cerebral c-fos and ornithine decarboxylase gene expression by trifluroperazine, *Eur. J. Pharmacol.*, 183, 1544–1545, 1990.

Zawia, N.H., and Harry, G.J., Developmental ODC gene expression in the rat cerebellum and neocortex, *Brain Res. Dev. Brain Res.*, 71, 53–57, 1993.

Zawia, N.H., and Harry, G.J., Exposure to lead acetate modulates the developmental expression of myelin genes in the rat frontal lobe, *Int. J. Dev. Neurosci.*, 13, 639–644, 1995.

Zawia, N.H., and Harry, G.J., Developmental exposure to Pb interferes with glial and neuronal differential gene expression in the rat cerebellum, *Toxicol. Appl. Pharmacol.*, 138, 43–47, 1996.

Zawia, N.H. et al., SP1 as a target site for metal-induced perturbations of transcriptional regulation of developmental brain gene expression, *Brain Res.*, 107, 291–298, 1998.

CHAPTER 2

Introduction to Molecular Biology and Its Techniques

Bingfang Yan

CONTENTS

041528031-1/04/$0.00+$1.50
11

2.1 CELLS, GENOMES, AND GENES

Cells are the basic units of structure and function for all living organisms. There are two types of cells: prokaryotic and eukaryotic. All prokaryotes, such as bacteria, are single-cell organisms, whereas all multi-cell organisms such as mammals are eukaryotes. One of the major differences between prokaryotic and eukaryotic cells is in the structures of membrane-restricted compartments. Prokaryotic cells lack a defined nucleus and have a single membrane-surrounded compartment. Eukaryotic cells, in contrast, have a defined nucleus and many organelles that are surrounded by a lipid membrane. These independent compartments or organelles in mammalian cells include the mitochondria, the endoplasmic reticula, the Golgi apparatus, the lysosomes, and the peroxisomes. The mitochondria are responsible for energy production; the endoplasmic reticula are involved in the synthesis of proteins and lipids and the metabolism of hormones and xenobiotics; the Golgi complexes direct protein and lipid trafficking; the lysosomes degrade worn-out cell constituents; and the peroxisomes are involved in lipid metabolism. Finally, the nucleus, although only loosely membrane-surrounded, contains the most publicized macromolecule deoxyribonucleic acid (DNA).

Although all somatic cells in a multi-cell organism contain similar basic structures such as organelles, they may differ significantly in terms of overall morphology and functional characteristics. Even in a case where cells have a similar histological appearance, they may perform a different function and be involved in different biological events.

The nervous system is a good example in this regard. Neurons and glia are two major cell types in the nervous system. The neurons are responsible for the tasks of the nervous system, whereas the glial cells occupy the spaces between neurons and modulate their functions. Trillions of neurons perform specialized functions, including storage, computation, integration, and transmission of information from sensory to motor stimuli.

All neurons contain four distinct structures: the cell body, the dendrites, the axon, and the axon terminals. The cell body contains the nucleus and is the site of biosynthesis (e.g., proteins and lipids). Also, the cell body is responsible for the degradation of worn-out neuronal constituents. The axon is a long process and transmits conduction of action potential away from the neuron body. The dendrites are smaller processes that receive signals from axons of other neurons.

The axon terminals are where neuronal signals are transmitted to other cells through a structure called the synapse. A synapse consists of a presynapse component (the axon terminal), a synaptic cleft, and a postsynapse structure. Generally, an action potential reaching an axonal terminal triggers the release of chemicals called neurotransmitters into the cleft. The diffused neurotransmitters act on the receptors of postsynaptic cells. Postsynaptic cells can be another neuron, a muscle, or a gland. Therefore, the transmitted signal can travel to another neuron or cause contraction of muscle, or it can initiate secretion in a gland. Conduction of action potential is an energy-driven process and requires good insulation (through the myelin sheath); therefore, neurons are highly sensitive to hypoxia, mitochondrial toxicants, and disruptors on neuronal cytoskeletons, as well as myelin integrity.

The basic compositions of DNA molecules are four nucleotides: adenine (A), guanine (G), cytosine (C), and thymine (T). The nucleotides form repeat units of a DNA chain through phosphodiester linkages. Nuclear DNA molecules are helical, non-circle and double-stranded. Two helical chains are held together through hydrogen bonds between A and T or C and G, based on Watson-Crick rules.

DNA molecules are associated with proteins to form chromatins. The most abundant proteins complexed with DNA are histone H1, H2A, H2B, H3, and H4. Every 146 base pairs of DNA encircle a core formed by two molecules of each of the last four-histone proteins. This DNA-wrapped histone structure is called a nucleosome and is the basic unit of the chromosome (Asturias, F.J. et al., 2002). In addition to histone proteins, there are many other proteins that are associated with DNA, notably scaffolds and transcription factors. Scaffold proteins provide anchoring supports for long chromatin loops, whereas transcription factors play an important role in the regulation of gene expression. Each eukaryotic chromosome contains a single linear DNA molecule approximately 1×10^8 base pairs long, depending on the length of a chromosome. The total number of chromosomes varies from species to species. For example, humans have 23 pairs, whereas mice have only 20 pairs of chromosomes. An entire set of chromosomes constitutes the genome. The human genome contains a total of 3×10^9 base pairs. In addition to chromosomal DNA, the mitochondria contain a circular DNA molecule, but it is only 1.6×10^4 base pairs long.

The most fascinating aspect of the DNA molecule is that it carries all genetic information. The completion of the human genome sequence as well as the near completion of the genomes from several model species such as the mouse have uncovered several general sequence patterns that define physical and functional units of heredity (Figure 2.1). These basic units are commonly called genes. Although the sequences of various genes are highly distinct, particularly in the transcribed region, there are several regions that are relatively conserved, particularly in the core promoter region.

The transcribed region acts as the template for the synthesis of the RNA molecule and starts immediately after a conserved sequence called the TATA box. This box and its adjacent sequence form the core promoter where the basic transcription machinery is bound. The region located immediately upstream of the core promoter

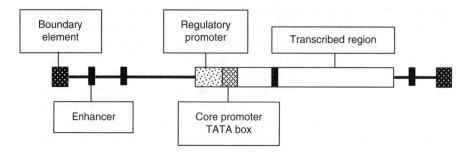

Figure 2.1   Structural features of genes.

is called the regulatory promoter and contains clustered elements where proteins bind and regulate the transcription rate of a gene. Farther away, either upstream or downstream from the core promoter, a gene may have several independent DNA elements, traditionally called enhancers (Chen, J.L. et al., 2002). These regulatory elements bind to a specific protein and coordinate the transcription of a gene (Blackwood, E.M., and Kadonaga, J.T., 1998).

Finally, all genes contain highly conserved regions called boundary elements that are located between two consecutive genes and act as insulators (Bell, A.C., West, A.G., and Felsenfeld, G., 2001). There are many exceptions regarding the role of putative boundary elements. In some cases, a promoter is flanked by two transcribed regions and controls the transcription rate of both genes. In other cases, a mini-gene is located within a larger gene and this mini-gene is called a "nested gene." Whether a nested gene has boundary elements and how these elements execute their activity remain to be elucidated. Some genes, structurally similar to functional genes, fail to support transcription or the transcribed mRNA species do not encode proteins. These non-functional genes are called pseudogenes. Pseudogenes are presumably derived from gene duplication, but contain many mutations that render inactivation of the promoter or cause frame-shift in the coding region.

Biological functions of genes are executed by proteins, not by DNA. The RNA molecule acts as a connector between a gene and the corresponding protein. The flow of genetic information from DNA to protein is called the central dogma of biology. It states that the coded genetic information hard-wired into DNA is transcribed into individual transportable cassettes, composed of messenger RNA (mRNA); each mRNA cassette contains the program for synthesis of a particular protein. Although the three macromolecules are present in all cells, the relative abundance varies markedly. DNA constitutes 1% of the total cell mass, RNA constitutes 6%, and proteins constitute 15%. Apparently, one DNA template makes several mRNA molecules, and one mRNA produces multiple protein molecules. The levels of mRNA and the corresponding protein collectively reflect the levels of gene expression. Chemically, RNA resembles DNA, with the deoxyribose being replaced by ribose in RNA. In addition, RNA contains the nucleotide uracil instead of thymine which is present in DNA. Although an RNA precursor is synthesized in the nucleus, mature RNA is located in the cytoplasm. In addition, mature mRNA differs from the precursor initially transcribed (Figure 2.2). Transcribed sequences present in the

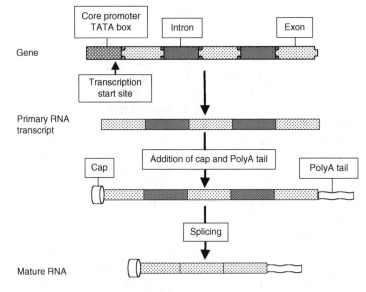

Figure 2.2 Synthesis and processing of RNA.

mature mRNA are called exons, whereas those absent in the mature mRNA are introns (Hamm, J., and Lamond, A.I., 1998). In addition, mature mRNA has several notable modifications such as the addition of a cap at the 5′ and a polyA tail at the 3′ end (Birse, C.E. et al., 1998; Sonenberg, N., and Gingras, A.C., 1998; Wahle, E., and Keller, W., 1996). These modifications are important for the stability of transcribed mRNA and for efficiency in directing protein synthesis.

The conversion process of mRNA into protein is called translation. An mRNA molecule can be divided into three parts, the 5′ and 3′ untranslated regions and the protein-coding region. The coding region (or open reading frame) can be specified in consecutive groups of three nucleotides. Each triplet, called a codon, recognizes a specific amino acid. All amino acids except methionine are assigned to more than one codon. UAA, UGA, and UAG specify no amino acid and constitute stop (terminator) signals. In addition to mRNA, there are two other types of RNA species involved in translation including ribosomal RNA (rRNA) and transfer RNA (tRNA). Different RNA species are synthesized by different RNA polymerases and have a different cellular abundance (Table 2.1). Ribosomal RNA molecules are associated with proteins to form ribosomes (including the large and small subunits) where translation takes place. tRNA carries amino acids and recognizes genetic codons.

Table 2.1   RNAs and Their Polymerases, Relative Abundance and Function

RNA	Polymerase	Relative Level (%)	Function
rRNA	I	80	Ribosomal assembling
mRNA	II	5	Protein coding
tRNA	III	15	Amino acid carrying

Structurally tRNA has an acceptor stem and an anticodon loop that are functionally important. The acceptor stem reacts with a designated amino acid to form aminoacyl-tRNA, which is catalyzed by a specific synthetase. The anticodon loop contains a three-base sequence that base-pairs with its complementary code in the mRNA. Each amino acid has its own tRNA, which is recognized by a specific aminoacyl-tRNA synthetase. With very few exceptions, translation starts at the codon that recognizes the first methionine and is surrounded by a so-called Kozak sequence (5'–ACC**AUG**G) (Kozak, M., 1997). The translation initiation complex, located at the first methionine codon, includes the large and small ribosomal subunits, an mRNA and Met–tRNA. Then translation proceeds by sliding on the mRNA stepwise over a distance of one triplet, with each slide incorporating a designated amino acid. The incoming amino acids are delivered by aminoacyl-tRNA that pair-matches with a slide codon on the mRNA through the anticodon triplet. The nascent polypeptide chain is terminated at a stop codon in the mRNA. Translation can occur with multiple ribosomes attached to a single mRNA. As a result, simultaneous syntheses of multiple polypeptide chains proceed. Generally translations in prokaryotic and eukaryotic cells are similar. However, transcription and translation are coupled processes in prokaryotic cells. In addition, prokaryotic cells may have different preferences on the selection of codons for a given amino acid. Therefore, some eukaryotic mRNAs may not be effectively translated in a prokaryotic expression system.

2.2 BASAL TRANSCRIPTIONAL MACHINERY

Gene expression is the entire process of gene-encoded information being converted into the corresponding protein. The levels of gene expression are usually determined through measuring steady levels of mRNA and protein. Factors that affect gene expression include: transcription rate, degradation of mRNA, efficiency of translation, and protein stability. Transcription determines whether and to what extent a gene is expressed; transcription rate therefore plays a determinant role in gene expression.

Transcription is initiated by the interactions of the basic transcription machinery with the core promoter of a gene (Lagrange, T. et al., 1998). For genes that contain a highly conserved TATA box (immediately upstream of the transcription start site), the interactions occur at the platform formed by this box and its adjacent 5'-flanking sequence. The basic transcription machinery contains RNA polymerase II (Pol II) and several general transcription factors (GTFs) including TFIIA, TFIIB, TFIID, TFIIE, TFIIF, and TFIIH. Most of the transcription factors are multimeric proteins with TFIID being the largest (~750 kDa). Although GTFs such as TFIIB recognize and bind to particular DNA sequences, only TFIID binds to a core promoter independently.

The basic transcription complex is formed through a stepwise assembling process. TFIID initially binds to the core promoter, followed by TFIIA, TFIIB, Pol II-TFIIF complex, TFIIE, and TFIIH. TFIIH has helicase activity; therefore the presence of TFIIH in the complex results in the melting of DNA (or the separation of double-stranded chains), which is essential for transcription. In the presence of ribonucleoside triphosphates, Pol II transcribes the template strand and slides away from the promoter (Geiduschek, E.P., 1998). Termination of transcription is achieved by a protein complex that cleaves and polyadenylates the 3′ end of a transcript, which occurs over a distance of 0.5–2.0 kb beyond the polyadenylation addition site.

Several points should be emphasized. Many eukaryotic genes lack the so-called TATA box; therefore, alternative mechanisms may exist. Although the basic transcription initiation complex is shown to confer transcription *in vitro*, it is not sufficient to execute transcription *in vivo*. Finally, the basic complex does not determine the extent that a gene is transcribed because this complex serves as a commonality for differentially transcribed genes.

Transcription activation or repression of a gene is achieved by specific proteins commonly referred to as transcription factors (contrasting to general transcription factors) (Maldonado, E., Hampsey, M., and Reinberg, D., 1999; Pabo, C.O., and Sauer, R.T., 1992; Torchia, J., Glass, C., and Rosenfeld, M.G., 1998). These factors have a DNA binding domain and one or more transcription regulatory domains (activation or repression). They bind to a specific *cis*-DNA element, recruit other proteins, and affect the performance of the basic transcription initiation complex. Some of the factors bind DNA as monomers, whereas others form homo- or heterodimers. In some cases, the dimerization status determines the binding affinity. For example, oncogenic proteins *c-fos* and *c-jun* heterodimers bind to DNA with high affinity, whereas *c-fos* homodimers bind to the same DNA element with only a modest affinity. The DNA binding domain is highly conserved and serves as a basis for classification of these factors.

Table 2.2 shows several major types of transcription factors, including homeodomain, leucine-zipper, basic helix-loop-helix, and zinc-finger. Homeodomain proteins contain two helix structures that are connected by a turn. The first helix interacts with DNA, whereas the second helix interacts with other protein(s) in the transcription machinery. Homeodomain factors are found to be important in regulating gene expression during embryonic development. Leucine zipper proteins contain several repeats of four or five leucines precisely seven amino acids apart. These repeats are hydrophobic and can interact with themselves or others to form dimers (homodimers or heterodimers). Leucine zipper proteins have an amino-terminus that is rich in lysine and arginine and mediates interaction with a DNA element. Leucine zipper transcription factors are exemplified by oncoproteins *c-fos* and *c-jun*. Basic helix-loop-helix proteins are similar to leucine zipper proteins in that they can form dimers and they have a positively charged region where DNA binding takes place. Many

Table 2.2 Eukaryotic Transcription Factors

Type	Structural Feature	Dimer for DNA Binding	Example
Homeodomain	Helix-turn-helix	Yes or no	PDX-1*
Zinc finger protein	Cysteine-histidine repeat	Yes or no	Steroid receptor
Leucine-zipper	Spaced leucine repeat	Yes	*c-fos* and *c-jun*
bHLH	Basic helix-loop-helix	Yes	DEC1**

* PDX-1, pancreas duodenum homeobox-1; **DEC1, differentially expressed in chondrocytes-1.

basic helix-loop-helix factors control cell differentiation and proliferation. The product of c-*myc* protooncogene is a basic helix-loop-helix protein that is deregulated or mutated in many malignancies. Zinc finger proteins contain repeated motifs of cysteine and histidine that fold up in a three-dimensional structure coordinated by a zinc ion. Based on the number of cysteines in the finger domain, transcription factors in this superfamily can be divided into C_2, C_4, and C_6 zinc finger proteins. $C_2 H_2$ zinc finger proteins such as Sp1 represent classic zinc finger proteins, and they are known to regulate the expression of genes related to cell growth and xenobiotic response. C_6 zinc finger proteins are represented by several well-characterized transcription factors in yeast such as GAL4, which is required for galactose metabolism.

C_4 zinc finger proteins constitute a large family of ligand-dependent nuclear receptors (Westin, S. et al., 1998). Notable examples in this family include steroid receptors, retinoid receptors, thyroid hormone receptors, peroxisome proliferate activated receptors, and many so-called orphan receptors such as the pregnane X receptor. Compared with $C_2 H_2$ finger proteins, C_4 proteins have several distinct characteristics. Some C_4 proteins are normally associated with cytosolic proteins such as heat shock proteins, reside in the cytoplasm, and bind to a ligand causing nuclear translocation. $C_2 H_2$ finger proteins have three or more repeating finger units, whereas C_4 finger proteins generally have only two repeating units. $C_2 H_2$ finger proteins generally bind to DNA as a monomer, whereas C_4 finger proteins bind to DNA as homo- or heterodimers. Even within the C_4 finger proteins, there are several notable differences. Some C_4 finger proteins such as the glucocorticoid receptor have a relatively deep ligand-binding pocket; therefore, they show a narrow spectrum of ligands and the binding is highly specific. In contrast, proteins in this family such as the pregnane X receptor have a rather large and flexible binding pocket, which allows them to interact with a vast array of compounds with dissimilar structures. In addition, these rather non-specific receptors are also shown to bind to varying DNA sequences. Therefore, these receptors provide important mechanisms in response to structurally diverse xenobiotics and have important toxicological implications.

2.3 REGULATION OF GENE EXPRESSION

DNA molecules in eukaryotic cells are associated with histone proteins, in a chromatinized form. The basic unit of chromatin is the nucleosome (Luger, K. et al., 1997), which is composed of ~146 base pairs of DNA wrapped tightly around a disk-shaped core of histone proteins. Such highly compacted structures prevent genes from being transcribed. Transcription activators and repressors have chromatin remodeling activity, modulate the tightness of the nucleosomes, and regulate the transcription of genes (Belotserkovskaya, R., and Berger, S.L., 1999). In addition, the physical and chemical state of a gene is a limited factor that modulates the interactions between histone proteins and DNA molecules (Gribnau, J. et al., 1998). Therefore, transcription control of eukaryotic genes is achieved through several distinct mechanisms: (1) the methylation state and the flanking sequences of a gene;

and (2) the concentrations and activities of activators and repressors (Jones, P.L. et al., 1998; Ng, H.H., and Bird, A., 1999).

It has been shown in many cases that transcriptionally inactive genes are highly methylated at the cytosine of CpG dinucleotides. In contrast, transcriptionally active genes are usually under-methylated. Although the precise mechanisms remain to be determined, it has been proposed that methylation prevents specific transcription activators from binding to *cis*-DNA elements or provides a platform for binding to methyl-CpG-binding proteins. In addition to methylation state, several specific DNA sequences, particularly so-called silencers, have been found to regulate gene transcription. These silencers are DNA elements that bind to silencing binding proteins. Such interactions direct the formation and maintenance of an inaccessible heterochromatin configuration. Heterochromatin represents highly condensed chromatins, and thus stains more darkly (Grunstein, M., 1998; Maison, C. et al., 2002). *In vitro* studies demonstrate that DNA in heterochromatin is less accessible to externally added proteins than DNA in euchromatin (the less condensed chromatin). In addition, sequences acting as locus control regions, insulators, matrix attachments, and even telomeres are involved in the formation of heterochromatin (Brown, K.E., 1997).

The crystal structure of the nucleosome reveals that the histone proteins have all arginines, face the surface, and interact directly with negatively charged backbone of the DNA molecule (Luger, K. et al., 1997). Therefore, acetylation of the arginine residues diminishes the positive charges and looses the contacts between histone proteins and DNA. Experimental evidence demonstrate that acetylation is associated with transcription activation, whereas deacetylation is associated with transcription repression. Acetylation and deacetylation of chromatins are collectively called chromatin remodeling (Pollard, K.J., and Peterson, C.L., 1998). Activators are usually present in a complex that has acetylase activity, whereas repressors are present in a complex that has deacetylase activity. Apparently activators or repressors likely direct the complex to a specific gene and a specific location through binding to *cis*-DNA elements (Chi, T. et al., 1995; Hanna-Rose, W., and Hansen, U., 1996). Several other proteins called chromatin-remodeling factors (CRF) are also involved in the modulation of chromatin. Instead of modulating the acetylation state of chromatin, CRFs have DNA and RNA helicase activity. Helicases disrupt interactions between base-paired nucleic acids as well as protein-nucleic acid interactions. The action of helicases is accompanied by ATP hydrolysis.

Given the importance of activators or repressors in the regulation of gene transcription, the concentrations and activity of these factors determine whether or to what extent a gene should be transcribed. In addition, the activity of an activator/repressor is dependent on post-translational modifications in some cases. For example, signal transduction activated transducers (STATs) represent a class of transcription factors that mediate cytokine signaling. They normally reside in the cytoplasm and are transcriptionally inactive. Following a cytokine stimulus such as interleukin-6, STAT proteins are rapidly phosphorylated, translocated into the nucleus, and transactivate their target genes.

Many genes require a set of transcription factors for transcription (Darnell, J.E., Jr., 1997; Cockell, M., and Gasser, S.M., 1999). For example, transcription of the transthyretin gene requires cooperative action among several transcription factors, including HNF1, HNF3, HNF4, C/EBP, and AP. The first two factors are expressed only in the liver, whereas the last three factors are present in many tissues. However, the transthyretin gene is transcribed only in the liver because of the absence of HNF1 and HNF3. In some cases, the overall activity of a transcription factor is dependent on its interacting protein. This is well exemplified by nuclear receptors. For example, heterodimers of retinoic X receptor (RXR) and thyroid hormone receptor (TR) bind to DNA and exert transcription repression activity. In the presence of a ligand (e.g., thyroid hormone), the heterodimers have transactivation activity. Further studies demonstrate that unliganded dimers bind to co-repressor NcoR, recruit deacetylase, and cause transcription repression. Interactions of the dimers with thyroid hormone result in changes of conformation. The altered confirmation causes the release of the corepressor accompanied by recruiting steroid coactivator-1, which subsequently causes transcription activation.

2.4 MONITORING OF GENE EXPRESSION

Gene expression is the entire process of genetic information (e.g., a gene) being converted into measurable phenotypes (e.g., a protein). The expression levels of a gene, however, are usually determined by measuring promoter activity, mRNA, and protein levels. Table 2.3 shows assays that are commonly used to determine gene expression. Some methods assay for only one of the three indicators, whereas others assay for more than one indicator. For example, Western analysis determines the levels of protein only, whereas an electrophoretic gel mobility assay is indicative of both promoter activity and protein levels in certain circumstances. Some approaches determine the overall expression of multiple genes, whereas others are used to measure the expression levels of a particular gene. For example, gene arrays can literally detect the expression of genes in an entire genome, whereas *in situ* hybridization is used to determine the expression of a single gene.

Table 2.3 Methods of Monitoring Gene Expression

Assay	Promoter Activity	mRNA	Protein	Analyzed Genes
Reporter	√			Single
DNase I footprinting	√			Single
EMSA*	√		√	Single
CHIP⸲	√			Multiple
Nuclear run-on	√	√		Single
Northern blotting		√		Single
DNA array		√		Multiple
RNase protection		√		Single
RT-PCR⸲		√		Single
Differential display		√		Multiple
In situ hybridization		√		Single
Western blotting			√	Single
2-D gel analysis			√	Multiple
Immunocytochemistry			√	Single

* EMSA, electrophoretic mobility shift assay; ⸲CHIP, chromatin immunoprecipitation; ⸲RT-PCR, reverse transcription coupled polymerase chain reaction.

Table 2.4   Comparison of Commonly Used Reporter Genes

Reporter Gene	Detection	Detection Limit (molecules 5x)
Chloramphenicol acetyltransferase	Chromatography	10^7
Luciferase*	Bioluminescence	10^5
β-galactosidase*	Colorimetric/chemiluminescence	10^5
Secreted alkaline phosphatase	Colorimetric/chemiluminescence	10^5
Human growth hormone	Radioimmunoassay	10^8
Green fluorescent protein*	Fluorescence	10^7

* Signals can be detected in cells.

2.4.1   Promoter Activity

Given the fact that the promoter of a gene dictates whether and to what extent a gene is transcribed, determination of the promoter activity is indicative of gene expression. One of the widely used assays to determine promoter activity is the reporter assay. A reporter is prepared by fusing the promoter of a gene to a sequence (usually called the reporter gene) that encodes a protein with readily measurable properties (Table 2.4). In most cases, a reporter gene encodes an enzyme such as luciferase. The reporter genes are usually derived from quite distanced species. For example, an insect luciferase gene is used for preparing a reporter construct containing an eukaryotic promoter. In most cases, reporter assays are performed by transient transfection experiments, and variation on the transfection efficiency between control and test wells may affect the measurement. To overcome such potential errors, a second reporter that does not respond to a test system is cotransfected to normalize transfection efficiency. In addition, a transiently introduced reporter may not reflect the expression of the corresponding endogenous gene. Therefore, in some cases, other assays are required to confirm results obtained from reporter assays. Nevertheless, reporter assays provide a quick and sensitive approach to determine promoter activity although the results are only indirectly indicative of gene expression.

2.4.2   DNase I Footprinting

Reporter assays are indicative of the overall transcription regulation of a gene: either activation or repression. However, they do not specify the sequence of the promoter that confers such a regulatory activity particularly when a large fragment of a promoter is assayed. As described above, transcription factors usually bind to a *cis*-DNA element and regulate the transcription of a gene. Therefore, identification of the protein binding region likely provides information on whether this region interacts with particular transcription factors. DNase I footprinting is an effective assay that is widely used for this purpose, particularly when protein binding sites have not been established for a gene. DNase I is an endonuclease that makes chain breakage nonspecifically, but preferentially at sites adjacent to pyrimidine nucleotides. However, such actions are applied only to naked DNA not to a protein-DNA complex. Therefore, a sequence in a promoter bound by a protein or proteins is protected from being cleaved.

In DNase I footprinting, a DNA fragment from a promoter is labeled at one end with ^{32}P. Part of the sample is digested directly with DNase I, the other part of the sample is incubated with nuclear proteins, and then subjected to DNase I digestion. Both reaction mixtures are then resolved by polyacrylamide gel electrophoresis. Autoradiographic bands absent in the reaction incubated with nuclear proteins suggest the presence of protein-DNA interactions. It is very important that the amount of DNase I used is limited so that each DNA molecule is cleaved only once.

2.4.3   Electrophoretic Mobility Shift Assay

The DNase I footprinting assay specifies the region to which a transcription factor binds. However, it cannot specify nucleotides that mediate such binding. In addition, DNA binding detected by DNase I is difficult to quantify. The electrophoretic gel mobility assay is an effective alternative for these purposes. This assay is based on the fact that protein-bound DNA has a lower electrophoretic mobility; therefore, it is shifted or retarded compared to unbound DNA. In a gel shift assay, DNA fragments, usually double stranded oligonucleotides, are radiolabeled and incubated with nuclear extracts for protein binding. The reactions are then resolved in an agarose gel. The protein-bound oligonucleotides migrate more slowly than unbound species. In some cases, an antibody specific to the bound protein is added to form even larger complexes (DNA–protein–antibody). The antibody-associated complexes have an even slower mobility than DNA–protein complexes, and thus cause a further shift or supershift. The addition of an antibody specifies the involvement of a specific transcription factor in binding to the radiolabeled DNA. Substitutions in the oligonucleotide can identify nucleotides that mediate the interactions. In addition, intensity of the autoradiographic band reflects the expression levels of the DNA binding protein, or, more precisely, the levels of the active form of the binding protein. This is of particular significance in cases where transcription factors require posttranslational modifications for DNA binding. For example, phosphorylated STAT proteins bind to their DNA elements, but native STAT proteins do not.

2.4.4   Chromatin Immunoprecipitation Assay

Gel mobility shift assay is performed with isolated nuclear extracts, and thus does not specify whether interactions between a transcription factor and DNA occur in the context of the genome. To determine whether interactions take place intracellularly, a chromatin immunoprecipitation assay is usually performed. In this system, chromatins in whole cells are subjected to cross-linking and then sonicated. DNA fragments directly bound to a transcription factor are precipitated with beads coated with an antibody against this factor. The precipitated DNA fragments are then analyzed for the presence of the promoter sequence of interest. In theory, chromatin immunoprecipitation can precipitate all promoter sequences that are bound by the transcription factor. Therefore, this assay allows DNA–protein interactions in multiple genes to be analyzed simultaneously.

2.4.5 Nuclear Run-On Assay

Chromatin immunoprecipitation assays specify the intracellular interactions between a transcription factor and its DNA element; however, they provide no information on whether such interactions confer transcription activation or repression. Although the reporter assay described above can differentiate between these opposing activities, determination of the transcription of a native gene is more definitive. For this purpose, nuclear run-on is usually performed. In nuclear run-on, nuclei are isolated and subjected to *in vitro* transcription in the presence of ^{32}P–UTP. The newly synthesized mRNA is radiolabeled, isolated, and hybridized to the corresponding cDNA, which is mobilized on a nitrocellulose or nylon membrane. The signals are visualized by autoradiography.

2.4.6 Determination of mRNA Levels

Levels of mRNA are frequently used as indicators for gene expression. Almost all assays monitoring mRNA levels are based on nucleic acid hybridization, and hybridization probes are readily prepared. Northern blotting electrophoretically separates all mRNA species, and thus specifies the molecular weights of the mRNAs analyzed. Array experiments monitor the levels of thousands of mRNA species simultaneously, literally allowing gene expression of the entire genome to be analyzed. RNase protection assays detect single nucleotide mutations, and are thus highly specific. Reverse transcription-coupled polymerase chain reaction experiments involve exponential amplification of an mRNA signal, and are thus highly sensitive.

2.4.7 Blotting Analyses

Northern blotting is the most common method for analyzing mRNA levels. In Northern blotting, total RNA or mRNA is electrophoretically separated in formaldehyde agarose gel. The samples are then transferred onto a nitrocellulose or nylon membrane. The blotted membrane is incubated in the presence of a probe specific to the mRNA species of interest. The probe is usually radiolabeled or labeled by a molecule that can be subsequently monitored. The hybridized mRNA with the probe can be visualized by autoradiography or other methods if a probe that is not radiolabeled is used. Another blotting analytic method is called slot or dot blotting, depending on the apparatus used. Instead of electrophoretic separation, total RNA or mRNA samples are blotted directly to a membrane sandwiched in a slot/dot apparatus. The blotted membrane is then detected with a probe as Northern blotting. Unlike Northern blotting, slot/dot blotting does not specify the molecular weights of the mRNA analyzed, therefore, it is not as specific as Northern blotting. However, slot/dot blotting is easier to perform and accommodates more samples than Northern blotting.

2.4.8   DNA Microarray

This is a recently established approach, largely due to the wealth of sequence information obtained through the human genome project and similar programs on several model species. Literally this method is capable of monitoring the expression of genes in a whole genome simultaneously (Luo, Z., and Geschwind, D.H., 2001). In this method, DNA fragments from individual genes are immobilized on closely spaced regions of a glass microscope slide. Based on the types of the immobilized DNA, there are two DNA microarrays: cDNA and oligonucleotide arrays. In the cDNA array, regions spanning ~1.0 kb are individually synthesized and mounted to the surface of a glass slide. Typically, an area (2 × 2 cm) can accommodate as many as 12,000 unique sequences. In the oligonucleotide array, DNA fragments with a length of 30 bp are chemically synthesized and immobilized to the glass slide. In some cases, the oligonucleotides are synthesized *in situ* on a solid support or chip. Apparently oligonucleotide arrays give more specific results. To detect the array slide, all mRNA species from one tissue or cell population are converted to cDNAs in the presence of fluorescent dye.

The fluorescence-labeled cDNAs are allowed to hybridize with the DNAs immobilized on the glass slide. The fluorescent signals are detected and reflect a pattern or profile of gene expression on a genomic scale. In cases where expression patterns from two different types of tissues or cells or the same tissues undergoing different treatments are compared, and the fluorescence-labeled cDNA probes are prepared with different fluorescent dyes, the ratios of the signals from one dye over those from the other are used for comparison on the expression profiles. Although DNA microarray allows the expression of numerous genes to be monitored simultaneously, the biological significance is rather difficult to draw. For practical reasons, results from array experiments usually provide initial leads for further studies.

2.4.9   RNase Protection Assay

Blotting analyses and array experiments are effective approaches for monitoring mRNA levels. However, these methods cannot differentiate between target sequences that are highly identical, particularly in cases where two sequences have only point or single codon mutations. In these situations, the RNase protection assay is usually performed. In this assay, an RNA probe is synthesized in the presence of a radiolabeled nucleotide(s). The radiolabeled probe is allowed to hybridize with target RNA species. The probe and mRNA hybrids are digested with RNase that cleaves single stranded RNA chains (mismatched regions). The digested samples are electrophoretically separated and detected by autoradiography. mRNAs that match perfectly to the probe yield a single band, whereas mRNAs that mismatch the probe yield multiple bands depending on the number of regions that mismatch the probe. This method detects a single nucleotide mismatching, and thus is highly specific. In addition, autoradiography is highly sensitive.

2.4.10 Reverse Transcription Coupled Polymerase Chain Reaction (RT-PCR)

The RNase protection assay provides approximately 10 times more sensitivity than Northern blotting. However, RT-PCR is even more sensitive. In theory it can detect a single transcript; therefore, it is particularly effective for studying genes that have a low expression level or gene expression in a single cell. RT-PCR is essentially the same as PCR, though the starting materials are different. RT-PCR starts with mRNA, whereas PCR starts with DNA. For RT-PCR, mRNA species are first converted into cDNA by a reverse transcriptase, and the synthesized cDNAs are then subjected to PCR amplification. PCR is a technique that exponentially amplifies a DNA template. In this method, a short oligonucleotide (primer) binds specifically to a region of a single stranded DNA. Such binding creates an initiation point for the elongation of the primer along with the DNA template (the primer bound DNA). The other strand of a duplexed DNA molecule can serve as a template as well, provided a primer is supplied to bind to it. The second primer usually binds to a region several hundred base pairs apart from the first primer and is extended from the opposite direction. As a result, the sequence spanned by two primers is amplified (Figure 2.3).

A typical PCR reaction requires only a tiny amount of DNA or cDNA species; dNTP; a thermostable DNA polymerase (e.g., *taq*); and primers that recognize the end sequences of the DNA to be amplified. Reaction tubes or plates are located in a thermocycler, which allows the temperatures to be changed in a period of seconds. Usually three temperatures are used, as described in Figure 2.3. A denaturing temperature (e.g., 95°C) is set to cause strand separation of a duplexed DNA molecule; an annealing temperature (e.g., 52°C) is set to allow the primers to bind to the separated DNA templates; and an extension temperature (72°C) is set to allow the primers to be elongated according to the template sequence. As a result, every temperature cycle leads to a 2× amplification of the templates. The temperature cycles are repeated many times so that low copies of templates can be detected. Many modifications, based on the basic protocol, are made to fit in specific situations. Particularly on the detection system, a quantitative PCR cycler allows the amplification process to be monitored so that a linear amplification range can be determined. Establishment of the linear ranges provides more accurate comparisons among samples from various treatments, which is particularly relevant to toxicological studies. Overall, RT-PCR, like PCR, provides a highly sensitive, rather simple, and fast method for determining mRNA levels.

2.4.11 Differential Display

This is a method for detecting response genes to treatments (e.g., toxic chemicals) or differentially expressed genes between two types of cells or tissues. It is similar to RT-PCR, but the primers used are much shorter and less specific. In a regular RT-PCR, a single RNA species is amplified, whereas in differential display, as many as

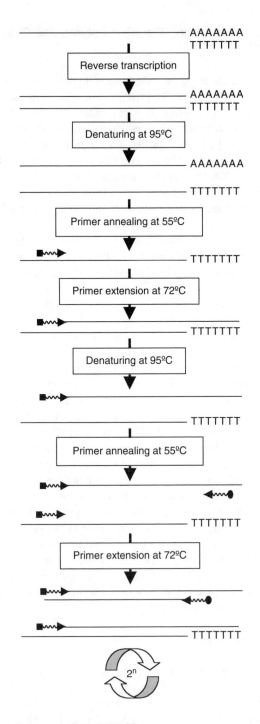

Figure 2.3 Schematic presentation of RT-PCR.

200 RNA species are amplified together. In differential display, reverse transcription is conducted with an anchored primer that recognizes polyA tails. During PCR amplification, this primer is coupled with another primer called arbitrary primer. This arbitrary primer is usually 12 base long and recognizes an arbitrarily defined sequence. Only the cDNAs with this arbitrarily defined sequence are amplified. In many cases, more than 10 arbitrary primers are used, with each of them being coupled with the anchored primer. Each coupled reaction analyzes ~200 mRNA species as visualized by sequence gel electrophoresis. Like DNA array experiment, mRNA differential display determines overall gene expression of given cells or tissues, and usually provides initial leads for further studies.

2.4.12 Determination of Protein Levels

Western analysis is by far the most often used approach for determining protein abundance. In Western analysis, proteins are denatured and resolved in an SDS polyacrylamide gel, transferred to a nitrocellulose or nylon membrane, and detected with an antibody. This antibody is conjugated with an enzyme, a fluorescence dye, or an isotope. These conjugated molecules are detected and the signals serve as indicators for protein abundance. Western analysis under SDS denaturing conditions specifies the size but not the charges of a protein. Many proteins, although translated from the same transcripts, may undergo post-translational modifications such as glycosylation and phosphorylation. Some of the modifications are highly associated with biological activity of a protein. Therefore, in some cases, it is desirable to differentiate the modified from the native form of a protein. Unfortunately, some post-translational modifications cause little change in the molecular weight; thus, regular Western blotting under the SDS denaturing condition may not differentiate modified from native forms of a protein. Alternatively, proteins are electrophoretically separated under nondenaturing conditions (without SDS), which allows proteins to be separated based on charges. Similarly, the separated proteins are transferred to the membrane and detected by an antibody.

To analyze proteins based on molecular weights and charges simultaneously, 2-D gel analysis is performed. Samples are initially subjected to an isoelectrophoretic focusing gel to separate proteins based on charges and then to a regular SDS electrophoresis to achieve separation based on protein sizes. Similarly, the SDS gel is blotted to a membrane and the proteins of interest are detected by an appropriate antibody. In most cases, 2-D gel separation is not detected with a specific antibody; instead, the detection is performed by measuring protein conjugates such as an isotope. In these cases, protein samples are radiolabeled before being subjected to isoelectrophoresis. A specific protein is identified based on its molecular weight and pre-established isoelectric point. In some cases, identification is made by picking up a protein spot and sequencing or mass spectrometry. Recently several non-charged fluorescence dyes are developed and used to label protein samples. Usually samples from control experiments are labeled with one dye, whereas samples from test experiments are labeled with another dye. The samples are then equally mixed and subjected to 2-D gel analysis. The ratios of signals from one dye over the other are used as indicators for the effect of the treatment on the gene expression. In summary,

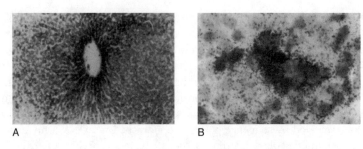

A   B

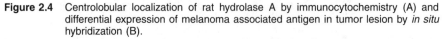

Figure 2.4   Centrolobular localization of rat hydrolase A by immunocytochemistry (A) and differential expression of melanoma associated antigen in tumor lesion by *in situ* hybridization (B).

2-D gel electrophoresis remains the method with the highest resolution power on protein separation and effectiveness on simultaneous analyses of multiple proteins, although so-called protein chip technology has rapidly developed in recent years (Wooley, J.C., 2002).

None of the techniques described above can specify the abundance of a protein in a specific subcellular compartment within a cell or a specific cell within a tissue. Immunocytochemistry provides an effective alternative in this regard. Typical immunocytochemistry studies involve fixation, embedding, sectioning, and immuno-detection. Fixation is usually achieved with a cross-linking agent such as formaldehyde; thus, fixation likely destroys epitopes that are recognized by antibodies, resulting in poor staining. Alternatively, tissue sections are prepared with a cryostat, which maximizes the native conformation of proteins. In addition, cryo-sectioning is a much quicker method. mRNA levels can also be detected in tissue sections. In this case, a nucleic acid probe is used, and this method is commonly referred to as *in situ* hybridization. Figure 2.4 shows the centrolobular localization of rat hydrolase A detected by immunocytochemistry and the differential expression of melanoma associated antigen by *in situ* hybridization. It should be emphasized that both immunocytochemistry and *in situ* hybridization are less quantitative compared with other methods such as Northern and Western analyses.

2.5 RECOMBINANT DNA TECHNOLOGY

Modern biotechnology is centered by recombinant DNA, which is defined as DNA fragments from two different sources ligated together and acting as a functional entity. One of the most exciting features regarding recombinant DNA is that almost any desirable mutation can be introduced into a recombinant DNA molecule. These mutations establish the importance of a particular sequence in the regulation of gene expression or of an amino acid in fulfilling the biological function of a protein. Preparation of recombinant DNA constructs usually involves the use of two types of enzymes: restriction endonucleases and ligases. Restriction enzymes recognize a specific sequence of a few nucleotides and make a cleavage without removing any nucleotides. In contrast, ligase joins two DNA fragments that have compatible ends (usually pretreated with the same restriction endonucleases). Another term that is

Table 2.5   Comparison of Genomic and cDNA Libraries

Feature	Genomic	cDNA
Initial materials	Genomic DNA	mRNA
Intron	Yes	No
Exon	Yes	Yes
Relative insert size	~20 kb	~1.5 kb
All somatic cells	Identical	Different
Expression of cloned genes*	No	Yes

* Depending on the host cells.

usually used exchangeably with recombinant DNA is molecular cloning, although molecular cloning is usually referred to as isolation of an unknown sequence (e.g., gene and cDNA).

2.5.1   Molecular Cloning

Molecular cloning starts with the preparation of gene libraries. There are two types of libraries with different starting materials. Libraries constructed with genomic DNA are called genomic libraries, whereas libraries constructed with cDNAs are called cDNA libraries. There are several major differences between genomic and cDNA libraries (Table 2.5). A genomic library contains both intron and exon sequences, whereas a cDNA library contains the sequences (e.g., exon) present only in the mature mRNAs. A genomic library contains rather large inserts (~50 kb), whereas a cDNA library contains much smaller inserts (~1.5 kb). Genomic libraries produce no recombinant protein, whereas cDNA libraries express proteins encoded by the inserts. Finally, genomic libraries from all somatic cells are the same, whereas cDNA libraries differ from cell to cell or tissue to tissue. Preparation of a genomic library involves four steps: isolation of genomic DNA, digestion with endonucleases to produce ligable DNA fragments, ligation of the digested DNA to a vector, and reconstitution of the ligated vector in host cells. In contrast, preparation of a cDNA library begins with isolation of polyA mRNA, followed by conversion of mRNA into cDNAs. The cDNAs are then used to prepare recombinant vectors followed by reconstitution in host cells.

2.5.2   Vector

By definition, vectors are carriers of cloned genes. Vectors have three essential features: (a) they must have a sequence that directs the replication of the vectors in host cells (e.g., bacteria); (b) they must have a sequence in which a cloned gene can be inserted (the sequence is usually called polylinker region); and (c) they must contain selectable markers. The markers serve two functions: to distinguish vector-containing host cells from those that do not contain vectors; to distinguish vectors that contain a cloned gene from those that do not. Commonly used vectors are plasmid, bacteriophage, cosmid, P1 vector, yeast artificial chromosome (YAC), and bacterial artificial chromosome (BAC). These vectors can replicate in bacteria or yeast, independently from the host chromosome. The plasmid is a relatively small

(~3.0 kb) vector, whereas the bacteriophage and P1 vectors are big (>40 kb). Cosmids contain elements from both plasmid and bacteriophage. The commonly used YAC vector contains a yeast centromere replication initiation region, two sets of telomere (ends of eukaryotic chromosome), selectable markers, and a cloning site. The BAC vector is similar to a regular plasmid vector with the exception that the BAC vector contains the origin, and genes encoding the ORI-binding proteins from a naturally occurring large *E. coli* plasmid called F-factor. Vector selection for library constructs is dependent on the sizes of inserts to be cloned and replication copies of a vector in host cells. Plasmid vectors are usually used for the construction of cDNA libraries, phage vectors are usually used for the construction of genomic libraries, and YAC and BAC are used to clone DNA fragments that are larger than 25 kb.

2.5.3 Library Screening

Two methods usually used to screen libraries are immunochemical approaches and nucleic acid hybridization techniques. Genomic libraries can be screened by only nucleic acid probes, because genomic inserts contain intron sequences that encode no proteins. In contrast, a cDNA library can be screened by both nucleic acid probes as well as antibodies. In both cases, a library is plated out, and individual colonies with an insert-containing vector are then transferred to a nylon membrane. After removing non-specific binding, the membrane is then incubated with a nucleic acid probe or an antibody. Nuclear acid probes are usually prepared by labeling a fragment of cDNA or an oligonucleotide. Labeling is performed with radioisotopes or other molecules such as biotin. Hybridization of recombinant vector DNA with radiolabeled probes is detected by autoradiography, whereas hybridization with non-radiolabeled probes is detected immunochemically. The colonies identified by hybridization or antibody are picked up and expanded. It should be emphasized that PCR has been increasingly used for molecular cloning due to the availability of abundant sequence information from many species. In addition, PCR is also used for library screening, particularly when the partial sequence of a gene or cDNA is known.

The positive clones identified by library screening need to be confirmed, usually by sequencing analyses. Two methods are used to determine DNA sequence: the Maxam-Gilbert method and the Sanger method. The Maxam-Gilbert method uses end-labeling and chemical cleaving approaches, whereas the Sanger method is the primary sequencing method and involves the use of chain terminators compose 2′3′-dideoxynucleoside triphosphates (ddNTP). These terminators can be incorporated normally into a growing DNA chain through their 5′ triphosphate groups. However, ddNTPs cannot form phosphodiester bonds with the next incoming deoxynucleotide triphosphates, because ddNTPs lack the 3′ hydroxyl group in the ribose, which is essential to the formation of phosphodiester bond. In the chain termination method, a small amount of a specific dideoxy NTP (i.e., ddATP) is included along with other dNTPs normally required in the reaction mixture for DNA synthesis by DNA polymerase. The ddATP molecules randomly incorporate into the chain during the elongation; therefore, the products are a series of chains with different lengths. However, all of the chains end with a terminator of ddATP. During the preparation of sequencing reactions, four separate tubes are used with each containing a different ddNTP. These reactions are then loaded into four consecutive lanes in a high-resolution polyacrylamide gel. The synthesized products are then visualized by staining or autoradiography. In the past several years, fluorescent dyes have been developed to label terminators with different colors; therefore, reactions containing all four terminators can be performed in one tube and analyzed in a single lane instead of four electrophoretic lanes.

2.5.4 Generation of Deletion and Point Mutants

Molecular cloning reveals detailed sequence information on a gene and its encoded protein. At the same time, sequence information can suggest structural features and potential functions of this protein. However, extensive biochemical and molecular experiments are required to definitively establish the functionality as well as the structure-function relationship. Mutants with deletions and substitutions of amino acids are widely used for these purposes. Deletion mutants have one or more fragments in a sequence being removed, whereas substitution mutants contain one or more nucleotides being altered. If the deleted sequence or substituted nucleotides in the coding region (sequence information for protein synthesis), the resultant proteins are altered. Deletion mutants are usually prepared with restriction endonu-cleases or PCR, whereas substitution mutants are usually prepared by oligonucle-otides. Figure 2.5 shows the schematic procedure to generate DEC-2 mutants that define the response element for DEC1-mediated regulation (Li, Y. et al., 2002).

DEC1 and DEC2 are two transcription factors that have highly identical protein sequences with an identical DNA binding motif. Both genes are inducibly regulated by many detrimental conditions such as hypoxia. In addition, DEC1 is found to be up-regulated in colon tumors and to have antiapoptotic activity against serum star-vation. Surprisingly, the expression of DEC2 is down-regulated in colon tumors. Reporter assays suggest that a 2-kb upstream sequence of the transcription initiation site in the DEC2 gene mediates suppression by DEC1. In order to determine the DEC1 response element within this 2-kb sequence, deletion and substitution mutants are prepared. The deletion mutants are prepared by PCR with primers that span sequences with a different length. The primers are extended to include restriction endonuclease sites to facilitate subsequent ligation. Studies with the deletion mutants identify the sequence (-150 to -170 from transcription starting site) that is required for DEC1-mediated suppression. This 21-bp sequence contains an E-box that is known to interact with the bHLH transcription factors. In order to establish defini-tively that this element confers DEC1-mediated suppressive activity, oligonucle-otides with substituted nucleotides [T(-161)G and C(-165)A] are used to prepare a mutant that disrupts this E-box. The mutant is prepared through oligonucleotides as described in Figure 2.5. Studies with the substitution mutant further establish that this element indeed is responsible for DEC1-mediated suppression. It should be emphasized that mutants are widely used to define functionally important residues and motifs in proteins as well.

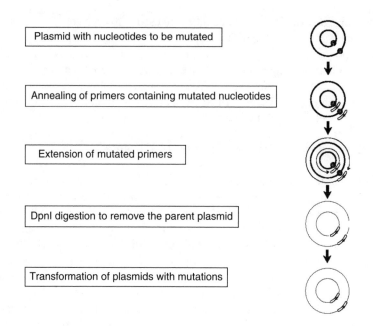

Plasmid with nucleotides to be mutated

Annealing of primers containing mutated nucleotides

Extension of mutated primers

DpnI digestion to remove the parent plasmid

Transformation of plasmids with mutations

Figure 2.5   Site-directed mutagenesis through oligonucleotides.

2.5.5   Production of Recombinant Proteins

One of the important applications of recombinant DNA is to produce recombinant proteins. Recombinant proteins can be made in such a way that some of the important amino acids are altered. As a result, the structure and function of a protein can be determined. Second, biologically active proteins can be made at large scales for industrial and medical purposes (Aebischer, P., and Ridet, J., 2001). Although proteins can be made in a heterologous system, their activity may not be as good as the native proteins. Most eukaryotic proteins require post-translational modifications and proper cell sorting. Some host cells used for the production of recombinant proteins may differ in processing them from their native host counterparts. In addition, recombinant genes are usually driven by a strong promoter. This is desirable because a large amount of protein can be produced; on the other hand, overproduction of a single protein may be beyond the host processing capacity.

There are five types of host cells that are commonly used for producing recombinant proteins: bacteria, yeast, insect cells, plant cells, and mammalian cells. Bacteria are prokaryotic, whereas the rest are eukaryotic. The *E. coli* expression system was the first technique developed; thus, the features of the hosts and the vectors used for expression are well established. In contrast, eukaryotic cells, particularly mammalian cells, are technically demanding, although these cells assure the production of functional proteins. The post-translational differences among these systems are summarized in Table 2.6.

Table 2.6 Comparison of Expression Systems on Post-Translational Modifications

Modifications	E. coli	Yeast	Mammalian Cells	Insect Cell
Proteolytic cleavage	+/–*	+/–	+	+
Glycosylation	–	+	+	+
Secretion	+/–	+	+	+
Folding	+/–	+/–	+	+
Phosphorylation	–	+	+	+
Acylation	–	+	+	+
Amidation	–	–	+	+
Yield (%)	~5	~1	>1	30

* A plus sign denotes yes whereas a minus sign denotes no.

2.5.6 Introduction of Recombinant DNA into Expression System

There are two terms that are usually used to describe introduction of a foreign gene into a host cell: transformation and transfection. Introduction of a gene into bacterial and yeast cells is called transformation, and it usually causes permanent and heritable alteration. In contrast, introduction of a gene into mammalian cells is usually called transfection. Transient transfection results in temporary expression of a gene, whereas stable transfection results in integration of the introduced DNA in the genome. The first step to perform transformation is to prepare competent cells that readily accept foreign DNA. Competent cells are usually prepared by treating highly proliferative cells (e.g., bacteria) with salts such as calcium chloride. Three methods are commonly used to transfect mammalian cells: precipitation, lipofection, and electroporation. The precipitation method is conducted by mixing the DNA molecules to be transfected with chemicals such as calcium phosphate. The DNA is precipitated and forms particles. These particles are taken up by host cells via phagocytosis. In lipofection, DNA to be transfected is mixed with lipids and then forms particles with the lipids covering the surfaces. These surface lipids fuse with host cell membrane. In electroporation, host cells are subjected to high voltage, which creates transient holes in cell membranes, through which the DNA to be transfected enters host cells. Almost in every case, particularly for stable transfection, DNA molecules to be introduced contain selective markers such as an antibiotic resistant gene for screening genomic integration of an introduced gene.

2.5.7 Purification of Recombinant Proteins

Like purification of native proteins, purification of recombinant proteins involves solubilization, selective precipitation, and chromatography. Generally, recombinant proteins are more abundant, and therefore they are relatively easy to purify. In addition, constructs encoding recombinant proteins are fused to a sequence encoding a tag that facilitates purification even more. The commonly used tags include 6xHis, glutathione S-transferase (GST), and maltose binding protein. The 6xHis tag contains 6 consecutive histidines, which have a high affinity toward nickel ion. GST is a phase II biotransformation enzyme that binds to glutathione. Maltose binding protein binds to maltose. These tags bind to respective target molecules reversibly, providing

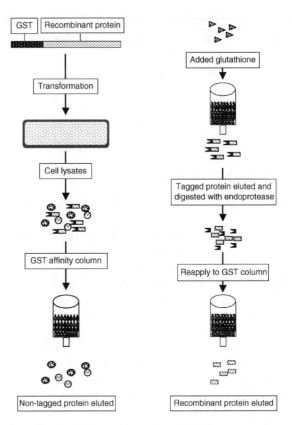

Figure 2.6   Affinity purification of tagged recombinant protein.

a tool for the purification of tagged recombinant proteins. Figure 2.6 shows the purification of a recombinant protein fused with GST. A construct is prepared by fusing the sequence encoding a protein of interest with the sequence encoding GST. This construct is used to transfect *E. coli* and cell lysates are prepared. The lysates are applied to a column packed with agarose beads coated with glutathione. The fusion proteins are retained in the column by binding to the glutathione beads, whereas other proteins are eluted. In order to elute the bound fusion proteins, free glutathione at high concentrations is added and displaces the bound fusion proteins. The fused GST can be removed through an endoprotease and the resultant digested mixtures are reapplied to the column to separate the recombinant proteins from the cleaved GST (Figure 2.6).

2.6 GENETICALLY ENGINEERED ANIMALS

Germline modifications of early embryos through recombinant DNA result in production of genetically altered animals. There are three types of modifications: insertion, deletion, and replacement. Insertion modification results in integration of a foreign gene or an additional copy of a native gene, and animals carrying additional gene are called transgenics. Deletion modifications selectively inactivate endogenous genes, and animals with disrupted genes are called knockouts. The third type of germline modification is knockin by which animals have a gene replaced by a foreign gene. Usually foreign genes used to replace endogenous genes are human homologues, genetic polymorphistic, or alternate splicing variants.

2.6.1 Transgenics

Transgenic animals are usually prepared using mice. This species provides advantages in breeding, housing, and vast existing scientific data. In addition, the fertilized eggs exhibit a clear nuclear boundary, which facilitates microinjection of foreign DNA. Injection is performed into one of the pronuclei of a fertilized egg. The injected DNA has a high likelihood of being randomly integrated into the chromosomes of the diploid zygote. The injected eggs are then transferred to foster mothers for embryonic development. Usually ~20% progenies contain the injected DNA, and pure transgenic strains are obtained through back-crossing breeding. The foreign genes carried by transgenics are called transgenes. Genomic integration through microinjection is very effective, as many as 100 copies of a transgene are usually present in transgenics. The expression of a transgene can be regulated in a temporal and tissue-specific pattern. The mechanism on forced expression of a transgene is based on the fact that many promoters have temporal and tissue-specific activity. For example, the insulin gene is driven by a pancreas-specific promoter; therefore, insulin is produced only in the pancreas. If a transgene is driven by the insulin promoter, the transgene will be expressed only in the pancreas. It should be emphasized that integration of a transgene is random; therefore, some integrations likely cause disruption of endogenous genes. For the same reason, transgenics harboring the same transgene but prepared at a different time or by other laboratories may behave differently in terms of biochemical and neurological phenotypes.

2.6.2 Knockouts and Knockins

In contrast to transgenics, knockout animals have one or more endogenous genes disrupted. Transgenics are initiated by microinjection of foreign DNA into fertilized eggs, whereas knockouts are started by transfecting embryonic stem cells (ES) with a recombinant gene-targeting vector. Instead of random integration, a construct contains part of the genomic sequence, which targets the construct to an endogenous gene of interest through homologous recombination (resulting in a disruption of this endogenous gene). In addition, the targeting construct contains two selectable markers: neo^r and th^{HSV}. The neo^r gene confers neomycin resistance, which screens for stably transfected cells. The th^{HSV} gene confers sensitivity to the cytotoxic agent ganciclovir, which screens against ES cells having nonhomologous recombination. ES cells with knockout mutations are injected into mouse blastocytes, which are subsequently transferred into a surrogate pseudopregnant mouse for embryo development. The resultant progenies, so-called chimeras, have tissues from both the transplanted ES and the host cells. These cells contribute to both somatic and germ cell population. Back-crossing breeding of the chimeras with the parent strain produces heterozygotes that carry the disrupted gene of interest. Further intercrossing of the heterozygous mice produces homozygous animals that have both copies of the gene disrupted. The same techniques can be used to generate knockin mice. Instead of inactivating a gene, knockin animals have a gene replaced by a foreign gene, in most cases by a human homologue.

2.6.3   Conditional Knockout

Complete inactivation of a gene provides a tool for defining the overall function of a gene. However, the function of this gene may differ from cell to cell, and from time to time. In the past several years, techniques have been developed to create knockouts in which a gene is inactivated in a temporal and/or tissue-specific manner. These genetically modified animals are called conditional knockouts. Overall, the protocol used to produce classic knockouts is applicable to the preparation of conditional knockouts. However, there is a major new development that makes the creation of a conditional knockout possible. The method is called Cre-loxP system. The Cre-loxP system is a cleavage and ligation system present in bacteriophage P1. Cre is a 38 kD recombinase that recognizes two loxP sites. A loxP site is a 34 bp sequence consisting of two 13 bp inverted repeats spaced by 8 nucleotides. If a DNA sequence contains two loxP sites, Cre-mediated recombination leads to a removal of the sequence between these two loxP sites. As a result, the gene involved is inactivated (Figure 2.7).

To develop conditional knockouts, two types of mice are required, with one expressing the Cre gene (Cre mouse) and the other having a DNA fragment flanked by two loxP sites (loxP mouse). The Cre mice are transgenics that express Cre recombinase. The transgene Cre is fused with different promoters so that all Cre

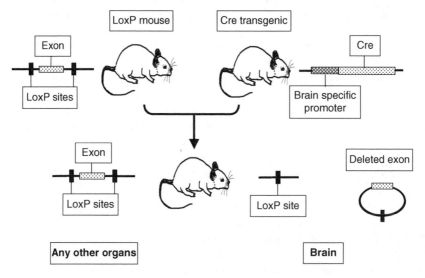

Figure 2.7   Conditional knockout through Cre-loxP system in the brain.

transgenics differ in terms of time and tissue for the expression of the Cre gene. The loxP mice will have a gene or part of the gene flanked by two loxP sites prepared by regular homologous recombination initiated in ES cells. These two loxP sites should be inserted into non-functioning sites of a gene so that the flanked gene has normal function. Through a cross-breeding between a loxP mouse and Cre mouse, the resultant progenies express Cre recombinase that causes the removal of the sequence between two loxP sites. As a result, this gene is inactivated. If the Cre gene is driven by a brain-specific promoter, the Cre-mediated recombination causes disruption of this gene in the brain. As described in Figure 2.7, a mouse has an exon of a gene flanked by two loxP sites, cross-breeding of this mouse with a mouse expressing the Cre gene in the brain causes a removal of this exon, thus specifically inactivating this gene in the brain but not other tissues (Figure 2.7). Conditional knockouts allow functions of genes to be determined in temporal and tissue-specific manner. In addition, conditional knockouts are particularly useful for genes that have embryonic lethal phenotypes for which classic knockouts cannot be developed.

2.6.4 Application of Genetically Modified Animals

Transgenic and knockout animals provide unprecedented means for all fronts of biology. These fronts include functional characterization of genes, development of disease models, and production of pharmaceutical and other biological agents. Many therapeutic products are manufactured in transgenic animals at a very large scale and a high purity. Molecular cloning has identified genes that are responsible for many genetic diseases. It is transgenic and knockout animals that provide the direct means to verify the importance of these genes. Confirmation of disease-related genes provides revolutionized tools for diagnosis and therapy of genetic-related diseases. With the completion of the human genome and the genomes from several model organisms, functional characterization of known and novel genes becomes the so-called postgenome focus of biological research. Knockouts and knockins under sophistic controls of gene expression allow the functions of genes to be established in a cell, tissue, or temporally restricted manner. Functional establishment of all genes will identify genes that are important for pharmacology (van der Neut, R., 1997). These so-called pharmacogenes will allow pharmaceutical companies to manufacture safer and highly selective therapeutic agents. Genetic analyses of these genes will likely provide an important step toward individual-based prescription or even individual-based manufacturing of medication. The excitement about the use and development of genetically modified animals has only been getting stronger since it started more than a decade ago. It is expected that many scientific ideas that are considered only fairy tales at the present time will be realized in the years to come.

2.6.5 Toxicological Models

Humanized animal models are increasingly used for drug-target identification, efficacy determination, and safety evaluation (Hakem, R., and Mak, T.W., 2001). Although model animals and humans share significant similarity on the genome level, there are many differences that determine species-specific effects. Transgenics, knockouts, and knockins provide effective tools to establish human relevancy and toxicological testing (Alden C., Smith, P., and Morton, D., 2002). For example, the cytochrome P450 system is involved in the metabolism of more than 95% of drugs and other xenobiotics. However, there are significant differences between human P450 and that of animals (e.g., the mouse). Mice that express human rather than mouse P450 enzymes generate more human relevant information. Genetically modified animals also provide tools to elucidate molecular mechanisms of many toxicants (Davidson, C. et al., 2001). Amphetamine and related compounds are powerful stimulants of the central nervous system. Overdose or chronic exposure causes symptoms similar to those in Parkinson's disease. Reactive oxygen and nitrogen species have been suggested to play determinant roles in amphetamine-induced neurotoxicity. Increased production of these species by amphetamine is likely achieved by interfering with the function of a dopamine transporter, triggering NMDA (N-methy-D-asparate) receptor-mediated signaling and subsequently activating neuronal nitric oxide synthase. In support of these mechanisms, knockouts of dopamine transporters or neuronal nitric oxide synthase are refractory to amphetamine-induced neurotoxicity (Gainetdinov, R.R., Mohn, A.R., and Caron, M.G., 2001). Similarly, transgenic mice over-expressing superoxide dismutase provide significant protection against amphetamine-induced neurotoxicity. These findings provide direct evidence that oxidative stress is a major mechanism for amphetamine-induced neurotoxicity.

2.6.6 Functional Characterization of Genes

Transgenics and knockouts provide an effective tool for defining the functions of a gene in whole animals. In the past decade, many genes are identified to be intimately associated with developmental events such as cell differentiation and lineage commitment. For example, basic helix-loop-helix (bHLH) proteins constitute a superfamily of well-characterized transcription factors. Activation and repression of transcription mediated by the bHLH proteins are physiologically important in organ development, and unbalance between these two mechanisms results in developmental defects. Mash1 bHLH protein heterodimerizes with E proteins and positively regulates neuronal differentiation. Mash1-null mice (mice with an absence of Mash 1) die at birth accompanied by loss of olfactory and autonomic neurons. HES proteins, mammalian homologues to *Drosophila* Hairy and E(spl), antagonize Mash1, Math1, and E proteins, thus negatively regulate neuronal differentiation. Absence of HES proteins (e.g., HES1) accelerates neuronal differentiation and results in severe defects in the nervous system such as anencephalic anomalies. HES1 was recently found to have a moderate proliferative effect. Differentiation inhibition and proliferation promotion featured by HES1 are thought to enable an increase in the number of committed cells necessary for full organ development.

Transgenics and knockouts allow functions of genes to be defined in a particular cell type or region in an organ. This is particularly of significance for the brain in which genes are expressed in a highly regionalized manner. Even within the same region, neurons in different subregions may perform unique biological tasks. For example, classic neuronal studies with lesion-induction, electrophysiological assays, and pharmacological reagents have established that NMDA receptors in the hippocampus are critical for spatial learning and memory. However, these method-ologies cannot specify whether NMDA receptors in the entire hippocampus or a subregion are required to confer such functions, largely due to difficulties in admin-istering pharmacological regents in a sub-region restricted manner. Conditional knockouts of the NMDA receptors through the Cre-loxP system have demonstrated the NMDA receptors in the subregion CA1 are responsible for spatial learning.

2.6.7 Disease Models

Many diseases occur due to genetic alteration. Some diseases are caused by inherited genotype, whereas others are caused by somatic mutations. In both cases, transgenic and knockout animals provide models to study pathogenesis and develop therapeutic strategies. In the past several years, many disease models have been developed through recombinant DNA technology, ranging from metabolic diseases to cancers and genetic diseases. Neurodegenerative diseases, commonly seen in the elderly, represent a major challenge to the scientific community and have been a major focus, largely due to our increasingly aging society. These diseases share a clinical characteristic, namely, the presence of insoluble aggregates in specific sub-sets of dying neurons. Molecular cloning has identified many genes that are linked to the development of these disorders, including Huntington's diseases, spinocere-bellar ataxias, spinal muscular atrophy, Parkinson's disease, and Alzheimer's disease. Transgenic models that partially or fully reproduce clinical features of these diseases have been developed.

Alzheimer's disease (AD) is the most dementing disorder of later life and a major cause of disability and death among the elderly (Dewachter, I. et al., 2000). Its clinical manifestations include memory loss, deficits in problem solving, and difficulty in language, with dementia being the diagnostic feature. Histological presentations include neurofibrillary tangles and senile plaques filled with deposited β-amyloid protein. Genetic studies have identified mutations that are linked to the development of AD. Notably the genes encoding amyloid precursor protein (APP), presenilin-1 (PS-1), presenilin-2 (PS-2), apoliproprotein E4 (ApoE4), and the protein *tau*. The protein *tau* stabilizes microtubuli as part of the axonal cytoskeleton and supports axonal transportation. Interestingly, the protein *tau* isolated from neu-rofibrillary tangles of AD patients is highly phosphorylated. Transgenic mice over-expressing human APP (mutants) develop part of the clinical manifestations of AD including amyloid plaques and cognitive and behavioral deficits. Double transgenic mice that coexpress APP and PS-1 show increased amyloid plaques that occur even earlier. Protein *tau* transgenic mice primarily exhibit psychomotorical impairment and axonopathy in the brain and spinal cord. ApoE4 transgenic mice, on the other hand, develop severe motoric symptoms and muscle wasting. In addition, drastic increases in phosphorylated *tau* protein and ubiquitin inclusions in hippocampal neurons are observed in ApoE4 transgenics. Overall, transgenics with these four genes manifest some important but not all AD symptoms, suggesting that AD is a polygenic disease. Nevertheless, studies with transgenic mice provide important information toward the elucidation of the molecular mechanisms involved in the development of AD.

REFERENCES

Aebischer, P., and Ridet, J. (2001), Recombinant proteins for neurodegenerative diseases: the delivery issue, *Trends Neurosci.* 24, 533–540.

Alden, C., Smith, P., and Morton, D. (2002), Application of genetically altered models as replacement for the lifetime mouse bioassay in pharmaceutical development, *Toxicol. Pathol.* 30, 135–138.

Asturias, F.J. et al. (2002), Structural analysis of the RSC chromatin-remodeling complex, *Proc. Natl. Acad. Sci. U.S.A.* 99, 13477–13480.

Bell, A.C., West, A.G., and Felsenfeld, G. (2001), Insulators and boundaries: versatile regulatory elements in the eukaryotic genome, *Science* 291, 447–450.

Belotserkovskaya, R., and Berger, S.L. (1999), Interplay between chromatin modifying and remodeling complexes in transcriptional regulation, *Crit. Rev. Eukaryot. Gene Expr.* 9, 221–230.

Birse, C.E. et al. (1998), Coupling termination of transcription to messenger RNA maturation in yeast, *Science* 280, 298–301.

Blackwood, E.M., and Kadonaga, J.T. (1998), Going the distance: a current view of enhancer action, *Science* 281, 61–63.

Brown, K.E. et al. (1997), Association of transcriptionally silent genes with Ikaros complexes at centromeric heterochromatin, *Cell* 91, 845–854.

Chen, J.L. et al. (2002), Enhancer action in trans is permitted throughout the *Drosophila* genome, *Proc. Natl. Acad. Sci. U.S.A.* 99, 3723–3728.

Chi, T. et al. (1995), A general mechanism for transcriptional synergy by eukaryotic activators, *Nature* 377, 254–257.

Cockell, M., and Gasser, S.M. (1999), Nuclear compartments and gene regulation, *Curr. Opin. Genet Dev.* 9, 199–205.

Darnell, J.E. Jr. (1997), STATs and gene regulation, *Science* 277, 1630–1635.

Davidson, C. et al. (2001), Methamphetamine neurotoxicity: necrotic and apoptotic mechanisms and relevance to human abuse and treatment, *Brain Res. Rev.* 36, 1–22.

Dewachter, I. et al. (2000), Modeling Alzheimer's disease in transgenic mice: effect of age and of presenilin1 on amyloid biochemistry and pathology in APP/London mice, *Exp. Gerontol.* 35, 831–841.

Gainetdinov, R.R., Mohn, A.R., and Caron, M.G. (2001), Genetic animal models: focus on schizophrenia, *Trends Neurosci.* 24, 527–533.

Geiduschek, E.P. (1998), Chromatin transcription: clearing the gridlock, *Curr. Biol.* 8, R373–375.

Gribnau, J. et al. (1998), Chromatin interaction mechanism of transcriptional control in vivo, *EMBO J.* 17, 6020–6027.

Grunstein, M. (1998), Yeast heterochromatin: regulation of its assembly and inheritance by histones, *Cell* 93, 325–328.

Hakem, R., and Mak, T.W. (2001), Animal models of tumor-suppressor genes, *Annu. Rev. Genet.* 35, 209–241.

Hamm, J., and Lamond, A.I. (1998), Spliceosome assembly: the unwinding role of DEAD-box proteins, *Curr. Biol.* 8, R532–534.

Han, J.H., Stratowa, C., and Rutter, W.J. (1987), Isolation of full-length putative rat lysophospholipase cDNA using improved methods for mRNA isolation and cDNA cloning, *Biochemistry* 26, 1617–1625.

Hanna-Rose, W., and Hansen, U. (1996), Active repression mechanisms of eukaryotic transcription repressors, *Trends Genet.* 12, 229–234.

Jones, P.L. et al. (1998), Methylated DNA and MeCP2 recruit histone deacetylase to repress transcription, *Nat. Genet.* 19, 187–191.

Kozak, M. (1997), Recognition of AUG and alternative initiator codons is augmented by G in position +4 but is not generally affected by the nucleotides in positions +5 and +6, *EMBO J.* 16, 2482–292.

Lagrange, T. et al. (1998), New core promoter element in RNA polymerase II-dependent transcription: sequence-specific DNA binding by transcription factor IIB, *Genes Dev.* 12, 34–44.

Li, Y. et al. (2002), DEC1/STRA13/ShARP2 is abundantly expressed in colon carcinoma, antagonizes serum deprivation-induced apoptosis and selectively inhibits the activation of procaspases, *Biochem J.* 367, 413–422.

Luger, K. et al. (1997), Crystal structure of the nucleosome core particle at 2.8 A resolution, *Nature* 389, 251–260.

Luo, Z., and Geschwind, D.H. (2001), Microarray applications in neuroscience, *Neurobiol. Dis.* 8, 183–193.

Maison, C. et al. (2002), Higher-order structure in pericentric heterochromatin involves a distinct pattern of histone modification and an RNA component, *Nat. Genet.* 30, 329–334.

Maldonado, E., Hampsey, M., and Reinberg, D. (1999), Repression: targeting the heart of the matter, *Cell* 99, 455–458.

Maxam, A.M., and Gilbert, W. (1980), Sequencing end-labeled DNA with base-specific chemical cleavages, *Methods Enzymol.* 65, 499–560.

Ng, H.H., and Bird, A. (1999), DNA methylation and chromatin modification, *Curr. Opin. Genet. Dev.* 9, 158–163.

Pabo, C.O., and Sauer, R.T. (1992), Transcription factors: structural families and principles of DNA recognition, *Annu. Rev. Biochem.* 61, 1053–1095.

Pollard, K.J., and Peterson, C.L. (1998), Chromatin remodeling: a marriage between two families? *Bioessays* 20, 771–780.

Sanger, F. (1981), Determination of nucleotide sequences in DNA, *Biosci. Rep.* 1, 3–18.

Shizuya, H., and Kouros-Mehr, H. (2001), The development and applications of the bacterial artificial chromosome cloning system, *Keio J. Med.* 50, 26–30.

Sonenberg, N., and Gingras, A.C. (1998), The mRNA 5′ cap-binding protein eIF4E and control of cell growth, *Curr. Opin. Cell Biol.* 10, 268–275.

Sternberg, N.L. (1992), Cloning high molecular weight DNA fragments by the bacteriophage P1 system, *Trends Genet.* 8, 11–16.

Torchia, J., Glass, C., and Rosenfeld, M.G. (1998), Co-activators and co-repressors in the integration of transcriptional responses, *Curr. Opin. Cell Biol.* 10, 373–383.

van der Neut, R. (1997), Targeted gene disruption: applications in neurobiology, *J. Neurosci. Methods* 71, 19–27.

Wahle, E., and Keller, W. (1996), The biochemistry of polyadenylation, *Trends Biochem. Sci.* 21, 247–250.

Westin, S. et al. (1998), Interactions controlling the assembly of nuclear-receptor heterodimers and co-activators, *Nature* 395, 199–202.

Wooley, J.C. (2002), The grand challenge: facing up to proteomics, *Trends Biotechnol.* 20, 316–317.

Yan, B. et al. (1994), Rat kidney carboxylesterase: cloning, sequencing, cellular localization and relationship to liver hydrolase B, *J. Biol. Chem.* 269, 29688–29696.

Young, R.A., and Davis, R.W. (1991), Gene isolation with lambda gt11 system, *Methods Enzymol.* 194, 230–238.

CHAPTER 3

Zinc Finger Transcription Factors Mediate Perturbations of Brain Gene Expression Elicited by Heavy Metals

Md. Riyaz Basha, Wei Wei, G. R. Reddy, and Nasser H. Zawia

CONTENTS

3.1 INTRODUCTION

Industrial and human activities have greatly increased the health risks associated with exposure to heavy metals. Transcriptional events are susceptible to environmental metals and their disturbance may result in perturbations in the regulation of gene expression. During the past decade, zinc finger proteins (ZFPs) have emerged as the largest class of transcription factors involved in the control of many aspects of life. Zinc fingers are structural motifs found in several classes of proteins proposed

041528031-1/04/$0.00+$1.50

to function in eukaryotic protein-nucleic acid interactions such as gene expression and DNA repair (Pabo and Sauer, 1992; O'Conner et al., 1993; Beckmann and Wilce, 1997).

Earlier studies on other organ systems and cells have demonstrated that proteins containing such motifs could be potential targets for perturbation by heavy metals. Recently, work by us and others have shown that metals such as lead (Pb) and mercury (Hg) interfered with the DNA binding properties of a member of this family of transcription factors, namely specificity protein one (Sp1). The most relevant aspect of these studies is the demonstration that the effects of these neurotoxic metals were mediated through interaction with the zinc finger domain of Sp1. However, Sp1 DNA binding is developmentally regulated and is inducible by growth factors. Therefore, cell signaling pathways are also involved in the maintenance and variation in ZFP DNA binding. This chapter will deal with the mechanisms involved in mediating the ability of neurotoxic heavy metals to induce structural and functional changes in these proteins, resulting in damage at multiple cellular sites.

3.2 TOXIC RESPONSES OF THE NERVOUS SYSTEM AFTER EXPOSURE TO HEAVY METALS

The neurotoxicants Pb and Hg affect many organs in the body; however, the developing brain is particularly susceptible to the neurotoxic effects of these metals (Rodier, 1986, 1990; Silbergeld, 1992; Clarkson, 1997). Chronic exposure to Pb results in alterations in axonal and synaptic elaboration (McCauley et al., 1982; Bull et al., 1983; Nichols and McLachlan, 1990). The mechanisms by which Pb alters neurodevelopment are not known; however, it is suggested that Pb can substitute for calcium and possibly for Zn in ion-dependent events at the synapse, such as calcium channels and N–methyl D–aspartate (NMDA) receptors, and is responsible for the observed impairment of various neurotransmitter systems and enzymes such as protein kinase C (PKC) (Markovac and Goldstein, 1988; Shao and Suszkiw, 1991; Guilarte et al., 1995; Hashemzadeh-Gargari and Guilarte, 1999). Pb has also been shown to disrupt developmental gene expression (Zawia and Harry, 1995, 1996; Zawia et al., 1998). The greatest concerns from mercury vapor exposure today are related to more subtle effects such as preclinical alterations in kidney function and behavioral and cognitive changes associated with damage to the central nervous system (Clarkson, 1997). On the other hand, methylmercury produces a neuropathology chiefly characterized by degeneration of cerebellar granule cells with preservation of Purkinje cells (Clarkson, 1997; Nagashima, 1997). Additionally, methylmercury also perturbs a number of cellular processes, which include astrocytic failure to maintain the composition of the extracellular fluid (Aschner, 1996). The biological targets of Hg have been more consistent with its position in the periodic table. Being in the same class as Zn and cadmium (Cd), Mercury (Hg) has a high affinity for ligands containing sulfhydryl groups (Komulainen and Bondy, 1987; Verity et al., 1994; Goyer, 1996) and it is particularly associated with specific metal-binding proteins such as the metallothioniens.

The clinical symptoms and pathological consequences of both Pb and Hg exposure are different; however, both appear to target the same family of transcription factors, namely, the ZFP. Consequently, this class of transcription factors may mediate perturbations of gene expression by these neurotoxic metals. However, while several metals can disturb the ZFP transcription factors, the final outcome may not be the same. While the effects of Pb on ZFP DNA binding have been correlated to disturbances in gene expression, it is not known whether Hg-induced perturbations in Sp1 DNA binding would result in a similar "productive effect." Furthermore, Hg and Pb distribution in the body is not identical and thus it is possible that they may act on different ZFP or the same ZFP but in different sites. This would lead to dissimilar target genes being modulated by these neurotoxic metals. Furthermore, it is important to note that these metals have short-term or immediate effects that are manifested through their interactions with various cellular proteins such as signal transduction intermediates, ion channels, enzymes, etc., and long-term delayed effects which are mediated through changes in gene expression. The signal transduction/transcription coupling pathways that are targeted by both metals may also vary.

3.3 METAL-INDUCED PERTURBATIONS IN THE Sp FAMILY OF TRANSCRIPTION FACTORS

Sp1 has a DNA binding domain composed of three zinc fingers, which are characterized by the cys2/his2 motifs (Kadonaga and Tjian, 1986; Kadonaga et al., 1987). These motifs consist of 26–30 amino acid residues, which are arranged in a semicircular structure that fits snugly into the major groove of B-DNA. Each zinc finger motif consists of an antiparallel β-hairpin and an α-helix, held together by a zinc ion and by a set of hydrophobic residues. Two thiol ligands of the cysteine residues from the β-sheet region and two N3 nitrogens of the imidazole side chains of the histidine from the α-helix coordinate the Zn ion in a tetrahedral structure (Lee et al., 1989). The α-helix of each zinc finger fits directly into the major groove of DNA and those residues from the NH2 terminal portion of each α-helix contact the base pairs in the major groove. Each of the three Sp1 zinc fingers uses its α-helix in a similar fashion, and each finger makes its primary contacts with a 3-bp subsite. The β-sheet, on the other hand, is on the back of the α-helix away from the base pairs and is shifted toward one side of the major groove. The two strands of the β-sheet have very different roles in the complex. The first β-strand does not make any contacts with the DNA, whereas the second β-strand contacts the sugar phosphate backbone along one strand of the DNA (Lee et al., 1989). Zn does not directly mediate protein–nucleic acid interactions, but instead, in combination with hydrophobic residues, it stabilizes the active conformation of the protein. This enables Sp1 to bind DNA, which is a prerequisite for this protein to participate in the transcriptional complex.

It has now become apparent that Sp1 belongs to a family of proteins whose members include Sp2, Sp3, and Sp4. All the Sp factors recognize the same DNA element. While the role of Sp2 is unknown, Sp3 is suspected of being a negative modulator of transcription and Sp4 exhibits tissue specificity for the brain (Suske, 1999). Sp1 activates transcription of viral and cellular genes that contain within their transcriptional control regions at least one Sp1 binding site consisting of a GC-rich decanucleotide sequence, and a GC box (Dynan and Tjian, 1983a,b; Gidoni et al., 1984, 1985). Some of the genes under Sp1 control include: ornithine decarboxylase (Briggs et al., 1986), myelin basic protein (MBP) (Gencic and Hudson, 1990), proteolipid protein (PLP), jc virus (Henson et al., 1992), simian virus 40 (SV40) (Dynan and Tjian, 1983a,b; Gidoni et al., 1984, 1985), NMDAR1 subunit (Bai and Kusiak, 1995), and metallothionein (Koizumi and Otsuka, 1994).

Considerable work has gone into examining the interactions between transition metals and zinc finger proteins. However, much of what has been done has concentrated on determining the structural flexibility of such motifs in the presence of different ions and not from the perspective of environmental health and toxicology. Fewer studies have concentrated on the potential of the disruption of these pervasive protein structures and carcinogenesis. Sunderman and Barber (1988) demonstrated that metals such as Cd and Ni could substitute for Zn within "finger loops" and potentially lead to carcinogenesis. Thiesen and Bach (1991) used rhSp1 as an apoprotein and reconstituted it with cadmium (Cd), cobalt (Co), copper (Cu), manganese (Mn) and nickel (Ni). They found that reconstitutions with Cd and Co were active, while those with Ni and Mn were less active. Following dialysis of the estrogen receptor and reconstitution with such metals, Predki and Sarkar (1992) reported similar findings. Thus it appeared that the zinc finger motif contained the ability to accommodate closely related ions with different radii. Consistent with the structural flexibility of these proteins, Fpg protein, a DNA-repair enzyme, also demonstrated the ability to incorporate Cd and Hg in its cysteine rich zinc finger-binding domain (O'Conner et al., 1993). While the above studies examined whole ZFP proteins, other researchers designed zinc finger peptides with predictable sequence specificity. Using NMR and gel mobility shift assays, these peptides were shown to assume a zinc finger formation when complexed with Zn. This demonstrated that DNA binding ability was subjected to displacement by other ions such as Cd and Co (Lee et al., 1992; Desjarlais and Berg, 1993; Krizek et al., 1993).

Pb and Zn are divalent cations known to compete at several physiological sites (Petit et al., 1983; Petit and LeBoutillier, 1986; Tomsig and Suszkiw, 1996). They also display similar affinities for metal-binding motifs present in calcium channels (Busselberg et al., 1991), neuronal NMDA receptors (Guilarte et al., 1995; Nihei and Guilarte, 1999), and PKC (Markovac and Goldstein, 1988, Tomsig and Suszkiw, 1996). Metal interactions with DNA binding proteins that regulate gene expression could modify protein conformation and thereby produce alterations in gene expression. Previous studies by us (Zawia and Harry, 1996; Harry et al., 1996; Zawia et al., 1998) have demonstrated the ability of Pb to selectively disrupt the gene expression of critical neuronal and glial genes such as growth associated protein 43 (GAP-43), MBP, and glial fibrillary acidic protein (GFAP). More recent work in our lab and others has found that Pb exposure alters the expression of NMDAR1 subunit (Brydie et al., 1999; Basha et al., 2003). The above genes contain response elements in their upstream regulatory portions that are recognized by ZFP, namely Sp1 (Henson et al., 1992; Bai and Kusiak, 1995).

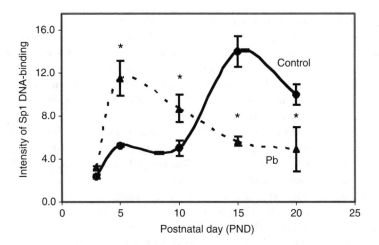

Figure 3.1   Changes in the developmental profiles of Sp1 DNA binding in the hippocampus
of control and Pb-exposed animals. Sp1 DNA binding was monitored using the
gel mobility shift assay. Shifted bands were scanned and quantitated using image
acquisition and analysis software (UVP, Inc., Upland, CA). Values are mean ±
S.E. of four independent experiments. Values marked with * are significantly
different from control (p <0.05) as determined by ANOVA.

In order to test whether Pb-induced alterations in gene expression occurred
through perturbations of ZFP, we examined changes in Sp1 and Egr-1 DNA binding
(Zawia et al., 1998; Reddy and Zawia, 2000). In general, we observed low Sp1 DNA
binding and Egr-1 DNA binding on the first week after birth, with a gradual increase
to peak around PND 15 in the control animals (Figure 3.1). In Pb-exposed animals,
a premature ZFP DNA binding appeared early on the first week with a drastic
decrease on subsequent days. More importantly, the pattern of changes in DNA
binding resembled the patterns of expression of target genes. Specifically, selective
premature peaks of MBP, PLP, and NMDAR1 mRNA expression were observed to
occur in a manner relative to the changes in Sp1 DNA binding (Zawia et al., 1998;
Brydie et al., 1999; Basha et al., 2003). Since these genes are frequent targets for
Sp1, these data suggested that exposure to heavy metals may alter developmental
gene expression and brain development through selective modulation of the transcriptional activity of Sp1. The developmental profile of non-ZFP did not exhibit
such changes (Zawia et al., 1998). Furthermore, Sp1 exhibited similar Pb-induced
changes in DNA binding in all brain regions, which clearly suggested that the
observed Pb-induced alterations may be specific for ZFP and are mediated by similar
mechanisms.

3.3.1   Direct Interactions by Heavy Metals at the Zinc Finger Domain

The addition of metals *in vitro* to nuclear extracts containing ZFP transcription
factors is one way to assess whether divalent metal cations can alter the DNA binding
of transcription factors. The above studies demonstrated that specific metals such as
Pb have the ability to selectively alter Sp1 DNA binding *in vivo*. However, the

mechanism by which this was occurring was not clear. One possible direct way through which the activity of ZFP may be modulated is by the interactions with the Zn binding sites in these motifs. Thus, we added a series of metals to nuclear extracts from HeLa cells and examined Sp1 DNA binding. Metals such as Pb, Zn, and Cd dramatically altered Sp1 DNA binding in a concentration-dependent manner, while Ca, Mg, and Ba had minimal to no effects on Sp1 DNA binding (Zawia et al., 1998). Unlike Zn, Pb is not a transition metal and therefore its effects may have been indirect and mediated by other cellular components present in the extract. To eliminate such a possibility, some of the above metals, as well as Sn (in the same class as Pb), were incubated with recombinant human Sp1 protein (rhSp1). Similar findings were again observed with rhSp1 as seen with the extracts; however, inhibition of DNA binding was observed at much lower concentration (Razmiafshari and Zawia, 2000). When group IIb metals such as Cd and Hg were incubated in the reaction medium, which includes rhSp1 and the labeled DNA consensus sequence, an abolishment of Sp1 DNA binding is observed at low concentrations (> 1.5 μM for Cd and > 5 μM for Hg). Again Pb exhibited a unique effect not seen with other non-transition metals, including its close neighbor Sn. Pb at concentrations greater than 37 μM eliminated Sp1 DNA binding. Furthermore, incubation of the DNA probe alone with these metals did not alter its mobility. It is possible that high Zn ion concentrations or the presence of additional xenobiotic metals interferes with the tetrahedral coordination of Zn in the zinc finger domain by altering its protein structure. Zn is critical for the binding of these transcription factors to DNA; therefore, the loss or absence of DNA binding suggests that these exogenous metals may have displaced Zn from its binding site leading to an alteration in the classical zinc finger structure required for DNA binding (Thiesen and Bach, 1991). The inability of Sn to modulate Sp1 DNA binding further reinforced the unique properties of Pb and its ability to directly alter Sp1 DNA binding by acting on the Sp1 protein.

While the above studies show that metals selectively alter Sp1 DNA binding by interacting directly with the Sp1 protein, it was important to see whether the effects of Pb are directed at the Zn binding site in the protein. A synthetic peptide composed of 26 amino acid residues that make up a zinc finger motif was synthesized (Razmiafshari and Zawia, 2000). The chemical and structural properties of this peptide complexed with various metals were then studied using light spectroscopy as well as DNA binding. Addition of either Pb or Hg to the synthetic apo-peptide generated the characteristic tetrahedral formation and thiolate coordination (Razmiafshari and Zawia, 2000). However, divalent metals such as Ca gave negative absorbances at the requisite wavelengths and failed to produce the characteristic pattern. An important question to address was whether those metals that were able to display the tetrahedral formation and cysteinate coordination of Zn are able to displace Zn from its binding site. The addition of Pb to Zn-peptide complexes appears to displace Zn and retain and enhance the signature spectrum of the Zn-peptide complex. Since the peptide synthesized is part of the DNA binding domain of Sp1, the next question to ask was, Do these peptide-metal complexes have biological activity? Can they show specific DNA binding properties? Using a highly modified and optimized gel mobility shift assay, the ability of these peptides and metal-peptide complexes to bind DNA was assessed. Consistent with the spectrophotometric studies, the metal–peptide complexes that did not produce a tetrahedral formation (Ca) also failed to exhibit any DNA binding to the consensus DNA element (Razmiafshari and Zawia, 2000). On the other hand, metals that exhibited zinc-finger-like spectra (Hg, Pb, Zn) were able to specifically bind the consensus oligonucleotide, in a concentration-dependent manner.

The metal-binding and structural properties of a synthetic peptide that resembled the zinc finger motif were investigated by one- and two-dimensional nuclear magnetic resonance (NMR). Group IIb metals Zn, Cd, and Hg and the non-transition metal Pb demonstrated distinct signal changes in the aliphatic region, which would allow some structural inferences such as metal-cysteine binding. Calcium, on the other hand, did not exhibit any significant chemical shift of proton resonances. All of the above metals, however, indicated major chemical shifts in the aromatic region of the spectra, which was indicative of metal-histidine binding. The chemical shift assignment, sequential connectivity and the interaction of the peptide with divalent metals, Zn, Pb, and Ca were monitored by 2D and the cross peak in TOCSY and NOESY spectra. Cysteine and histidine residues showed a distinct change in the amide and beta resonances region in the presence of Zn, Pb, and Hg, indicating the metal ligand-binding sites are near these residues. However, Ca was not actively involved in the binding site (Razmiafshari et al., 2001). Thus Pb and Hg are able to mimic the binding of Zn to specific residues and form a finger conformation (Figure 3.2).

3.3.2 Indirect Perturbations of Sp1 by Metals via Cell Signaling Pathways

We have reported that Pb and Hg directly interfered with the DNA binding of rhSp1, and Pb has been shown by others to alter the DNA binding of transcription factor IIIA (TFIIIA), implicating in both cases the zinc finger domain. Sp1 DNA binding is also developmentally regulated and is elevated in response to growth factors in vitro. The very strong association between high Sp1 expression and onset of differentiation is supported by coincidence of high levels of Sp1 in newly differentiating cells and relatively low amounts of Sp1 in fully differentiated cells with specialized functions (Saffer et al., 1991). The interactions of Sp1 with cell cycle regulatory proteins, including p53, retinoblastoma protein, and cyclins are consistent with its developmental role (Borellini and Glazer, 1993; Uvadia et al., 1993; Persengiev et al., 1996). Furthermore, Sp1 has been shown to play a critical role during the differentiation of oligodendrocytes in the human brain (Henson et al., 1992). While it is not known whether stimulation of Sp1 results in the activation of target genes that initiate the process of differentiation, it is evident that this transcription factor is highly responsive to neurotrophic and growth promoting signals since stimulation with nerve growth factor (NGF) alone increases Sp1 DNA binding activity (Zawia et al., 1998). The question of whether Pb exerts effects on differentiation by altering the activity and function of Sp1 is uncertain; however, it is believed that this transcription factor plays a critical role during early cell differentiation (Marin et al., 1997).

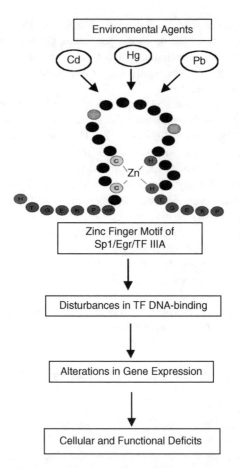

Figure 3.2   Putative pathway by which xenobiotic metals can interfere with ZFP-driven gene
expression. This pathway illustrates the direct action of xenobiotic metals on the
zinc finger motif of proteins, which can lead to structural alterations that result in
functional perturbations. For such a scheme to occur the environmental metals
must have the property to fit and coordinate with cysteine and histidine residues
in the metal coordination sphere of the protein. Sp1, Egr-1, and TFIIIA are used
as examples of transcription factors that could be subject to such modulation.

Therefore, we expect that Pb interferes with neural development by targeting
cellular events during the initiation phase of the differentiation process. We have
assessed the impact of Pb exposure on cellular differentiation by examining Sp1
DNA binding activity and neurite outgrowth, a morphological endpoint of differentiation in cultured pheochromocytoma (PC12). This cell line has been used extensively to study neurotrophic factor signaling and various downstream molecules that
are important in regulating proliferation and differentiation (Tischler and Greene,
1975). When PC12 cells are treated with NGF, they express a sympathetic neuronal-
like phenotype that is characterized by the cessation of proliferation and the promotion of neurite outgrowth (Greene and Shooter, 1980; Guroff, 1985).

In NGF-stimulated PC12 cells, Sp1 DNA binding activity was induced within 48 hours of exposure to NGF naive cells. Exposure of undifferentiated PC12 cells to Pb alone (0.1 μM) also produced a similar increase in Sp1 DNA binding. Since Pb altered the DNA binding profile of Sp1 in newly differentiating cells, neurite outgrowth was assessed as a morphological marker of differentiation to determine whether or not the effects of Pb on differentiation were restricted to the initiation phase (unprimed) or the elaboration phase of this process (NGF-primed). NGF-primed and unprimed PC12 cells were prepared for bioassay following exposure to various concentrations of NGF and/or Pb (Crumpton et al., 2001). Neurite outgrowth was measured at 48 and 72 hours during early stages of NGF-induced differentiation and at 14 hours in NGF primed/replated cells. In the absence of NGF, exposure to Pb alone (0.025, 0.05, 0.1 μM) promoted measurable neurite outgrowth in unprimed PC12 cells at 48 and 72 hours. A similar phenomenon was also observed in primed/replated PC12 cells at 14 hours. However, the effect was 2 to 5 times greater in unprimed cells. In the presence of NGF, a similar trend was apparent at lower concentrations, although the magnitude and temporal nature was different from Pb alone. In most cases, the administration of higher Pb concentrations (1 and 10 μM), in both the absence or presence of NGF, was less effective than the lower concentrations in potentiating neurite outgrowth. These results suggest that Pb alone at low doses may initiate premature stimulation of morphological differentiation that may be related to Pb-induced alterations in Sp1 binding to DNA (Crumpton et al., 2001).

There is biological plausibility to this convergence of Pb on NGF-stimulated differentiation. Previous evidence suggests that Pb may have effects on pivotal signal transduction events involved in cell differentiation like calcium mediated signaling of protein kinase C (PKC), which is also integral to NGF-stimulated differentiation (Laterra et al., 1992). Also, there is the suggestion that Pb's effects are concentration, time, and cell type specific (Kern and Audesirk, 1995). This explanation has been based in part on the known dose, time, and cell specific effects of agonists and antagonists of PKC, Ca, CaM kinase II, and PKA (reviewed by Kern and Audesirk, 1995). Previous *in vitro* studies with Pb on purified PKC from rat brain and neuronal cell cultures showed that Pb stimulated PKC activity at low doses and inhibited activity at higher doses, again suggestive of an inverted dose response curve (Mundy et al., 1996; Tomsig and Suszkiw, 1996). While the focus of previous mechanistic investigations of the effects of Pb at the cellular and molecular level have focussed on calcium-related events; to date no work to our knowledge has examined possible interactions of Pb with neurotrophic factor-mediated neuronal differentiation and its relation to zinc-finger proteins (e.g., Sp1). The novelty of this approach is that transcription factor activity may provide mechanistic information to support previous studies that examined morphological endpoints of differentiation in a number of cell types (Laterra et al., 1992; Kern and Audesirk, 1995).

The effects of methylmercury or mercuric chloride on neurite outgrowth and cell viability were quantified using undifferentiated (unprimed) and differentiated (primed) PC12 cells (Mundy et al., 2001). In unprimed cells, both methylmercury and mercuric chloride significantly decreased NGF-stimulated neurite outgrowth at concentrations of 0.3–3 μM. These effects were observed at 30-fold lower concentrations of Hg than those known to be cytotoxic (Mundy et al., 2001). These selective effects of Hg are morphologically different than those observed for Pb. While Pb promoted neurite outgrowth at low concentrations and inhibited them at high concentrations, the effects of Hg are a generalized inhibition. However, no correlations with ZFP DNA binding were done in the Hg study. Thus the relationship between neurite outgrowth and Sp1 DNA binding is not known. Hence, while both neurotoxic compounds may target ZFP, the outcome may be quite different.

3.3.3 Metal-Induced Disturbances in a Sp1-Driven Promoter

The promoter regions of genes contain the consensus elements that are recognized by transcription factors. One way to study the properties of these promoters and the requirements for their activation is to utilize constructs of such promoters linked to a reporter gene and to monitor reporter activity as an indirect measure of promoter function. The SV40 promoter contains six consensus elements for Sp1 and is linked to a luciferase reporter gene in a plasmid form (pGL3). This vector was cotransfected with pRLTK, whose reporter gene is driven by the thymidine kinase (TK) promoter, unrecognized by ZFP. The activity of TK serves as a control for transfection efficiency and basal transcriptional activity. PC12 cells serve as a host system for this transcriptional machinery. These cells express low levels of ZFP such as Sp1 whose DNA binding is dramatically elevated upon exposure to NGF. This DNA binding was also found to be modulated by the exposure to the neurotoxic metal Pb (see above). Thus a transcriptional machinery, monitored independently from that of the cell, was introduced into a cell in which the levels of ZFP can be manipulated by external factors.

The relative activities of the SV40 promoter were dramatically altered 4 hours following exposure to NGF and thereafter, while responses of the TK promoter were relatively unchanged (Figure 3.3). Addition of 0.1 μM Pb maximally induced SV40 promoter activity at 48 hours, while its combination with NGF lowered NGF-induced promoter activity. These results were similar to the effects of NGF and Pb on Sp1 DNA binding and differentiation observed in untransfected PC12 cells. We further studied the effects of Hg (mercuric chloride) and Ba (barium acetate) on this system. If our hypothesis was correct we expected that Hg would have an effect on a Sp1-driven transcriptional apparatus, while Ba would serve as a control. We found a dose-response relationship that was statistically significant, between the addition of Hg and promoter activity, while Ba had no effect even at concentrations of 10 μM. This was again consistent with our reported studies on the effects of these metals on the DNA binding of the Sp1 (Razmiafshari and Zawia, 2000).

Metal-induced changes in the activity of Sp1 may be consequent to alterations in intermediates in signal transduction pathways associated with growth. Studies from our lab had shown that Pb can cause the differentiation of PC12 cells and the induction of Sp1 activity in a manner similar to NGF (Crumpton et al., 2001). These findings suggest that neurotoxic metals could perturb developmental events through the involvement of the mitogen-activated protein kinase (MAPK) growth signaling pathway. To determine the relationship between Sp1 DNA binding, and MAPK,

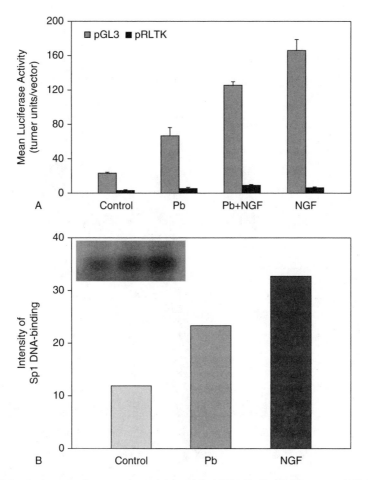

Figure 3.3 Sp1 responsive promoter activity and Sp1 DNA binding in Pb-exposed PC12 Cells. (a) Sp1 responsive promoter activity; (b) Sp1 DNA binding. PC12 cells were exposed to Pb (0.1 μM) and NGF (50 ng/ml) for 48 hours. pGL3 (Sp1-responsive) activity vs. pRLTK (Sp1-unresponsive) activity is plotted. Promoter activity values shown are the mean ± S.E. (n = 5–6). The Sp1 DNA binding from a representative experiment is also shown.

PC12 cells were exposed to Pb (0.025, 0.1, and 1 μM), NGF (0.3, 3, and 50 ng/ml) or a combination of both (0.1 μM Pb and 50 ng/ml NGF). The ability of the MAPK inhibitor PD 98059 (MEKI) to modulate Sp1 DNA binding activity was tested. Both Pb and NGF increased Sp1 DNA binding in a dose-dependent manner. MEKI dramatically reduced both basal and induced Sp1 DNA binding (Figure 3.4). These findings indicate that Pb's effects on Sp1 activity converge on a common pathway involving MAPK. These data suggest that stimulation of growth factor receptors appears to converge on the MAPK cascade potentially culminating in the regulation of nuclear transcription factors such as Sp1. Sp1 in turn may activate genes associated with growth and differentiation (Figure 3.5).

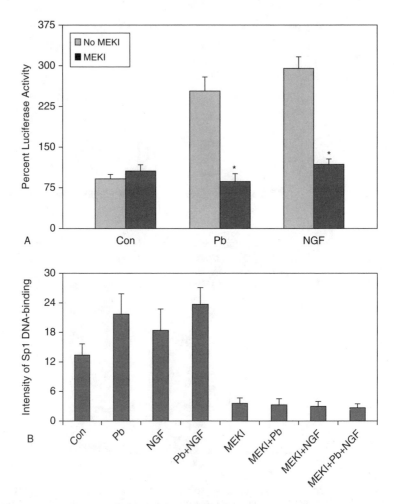

Figure 3.4 Effects of the MAPK inhibitor (MEKI) PD 98059 on SV40 promoter activity and
Sp1 DNA binding by Pb and NGF. SV40 promoter activity (a) and Sp1 DNA binding
(b) were measured 48 hours following exposure to Pb and/or NGF. The MAPK
inhibitor was added to some of the cells for the same period. Data shown represent
the mean ± S.E. (n = 3). Values marked with * are significantly different from control
($p < 0.05$) as determined by ANOVA.

3.4 THE Egr FAMILY OF TRANSCRIPTION FACTORS AND HEAVY METALS

Immediate early genes (IEG) are the first gene targets activated by the diverse
intracellular messenger systems linking membrane events and the nucleus (Beckman
and Wilce, 1997). The best characterized immediate early gene-encoded transcription
factors are the members of the Egr family. These factors were first identified about
a decade ago by several laboratories searching for genes whose expression was
induced by growth factors (O'Donovan et al., 1999). As a result, the four members

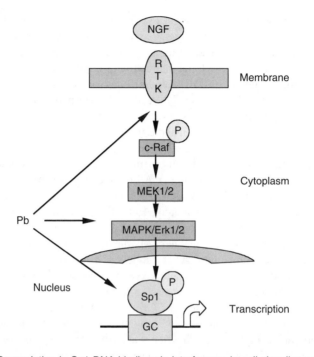

Figure 3.5   Deregulation in Sp1 DNA binding via interference in cell signaling pathways. The
functional activity of Sp1 may be modulated indirectly through interference with
cell signaling pathways associated with growth. The figure cartooned above (compiled from the literature) depicts how Pb may alter the regulation of gene expression
by interfering with signal transduction/transcription coupling via disruption of signaling intermediates (kinases). This is supported by studies that show MAPK
inhibitors (PD98059, U0126) abolish Sp1 DNA binding (see section on indirect
studies). NGF = nerve growth factor; RTK = receptor tyrosine kinase protein; c-raf
= serine/threonine kinase; MEK1 = MAPK kinase; MAPK = mitogen activated
protein kinase; Erk1 = extracellular signal-regulated kinase 1; GC = GC box
consensus element.

of this family were given multiple individual names. Egr-1, the prototypical member
of Egr family is also termed as zif268 (Christy et al., 1988), NGFI-A (Milbrandt,
1987), Krox-24 (Lemaire et al., 1988), and TIS8 (Lim et al., 1987) is a notable
example of IEG that is developmentally regulated in mammalian and avian brains
(Worley et al., 1991). Egr-1 was first isolated as NGFI-A by differential hybridization
from NGF treated rat PC12 cells, and has gained prominence as being one of the
most sensitive and specific markers of synaptic activity and plasticity (Yamagata
et al., 1994). Egr-1 encodes a transcription factor containing a DNA binding domain
formed by three zinc finger motifs (Rolli et al., 1999). The zinc fingers are arranged
in a semicircular (C-shaped) structure that fits snugly into the major groove of
B-DNA. Each zinc finger domain consists of antiparallel β-sheet and an α-helix
held together by a Zn ion coordinated by two cysteines from the β-sheet and two
histidines from the α-helix (Pavletich and Pabo, 1991).

Egr-1 binds to a separate GC-rich sequence in the promoter region of many
genes to regulate the expression of the target genes including growth factors and

cytokines. Egr-1 mRNA is expressed at low levels in the early postnatal rat cortex, midbrain and cerebellum, and the message increases throughout postnatal development to adult levels (Watson and Milbrandt, 1990). The transcriptional activity of Egr-1 could be regulated through the interaction of other transcription factors at the site of DNA-protein interaction. DNA binding sites for Sp1 and Egr-1 often overlap. Sp1 can bind to the Egr-1 consensus DNA motif (Ackerman et al., 1991; Ebert and Wong, 1995). The interactions between Egr-1 and Sp1 have been studied in detail whereas studies on how Egr-2, Egr-3, and Egr-4 interact with Sp1 and other members of Sp family, Sp2, Sp3, and Sp4 are not available. It was reported that no homology existed between the different Egr family members outside the DNA binding domain (Crosby et al., 1991; Joseph et al., 1988). All the members of the Egr family recognize the consensus motif GCG(G/T)GGGCG which is commonly referred to as GSG motif (Cao et al., 1990; Patwardhan et al., 1991; Crosby et al., 1991). Owing to the highly conserved nature of the zinc finger region among these proteins, all are likely to bind the DNA in a similar manner. All the Egr proteins show their highest expression in the striatum and CA1-3 subfields of hippocampus (Beckman and Wilce, 1997). Modulation of Egr-1 alters synapsin1 gene expression (Thiel et al., 1994) whereas alteration in Egr-2 expression results in abnormal hind-brain development (Dolle et al., 1993; Swiatek and Gridley, 1993).

The genes encoding Egr proteins exist in the mammalian genome as single copy genes and contain a single exon (Chavrier et al., 1989; Joseph et al., 1988; Patwardhan et al., 1991; Sakamoto et al., 1991). The intron, which is positioned 5' of the region encoding the zinc fingers, does not separate the protein into any recognizable domains. This means the zinc fingers are all encoded by the same exon in contrast to the gene encoding TFIIIA, where the first six of nine zinc fingers are each encoded on separate exons (Tso et al., 1986).

Much information is known about Egr-1 both in terms of its induction in response to extracellular stimuli and the molecular process governing its regulation and function (Beckman and Wilce, 1997). In contrast, relatively little is known about Egr-2 and even less about Egr-3 and Egr-4. Egr-1 is rapidly induced by growth factors and differentiation signals, including NGF and serum (Ackerman et al., 1991; DeFranco et al., 1993; Gaiddon et al., 1996; Kumahara et al., 1999), DNA damaging agents (Ahmed et al., 1997) and stress treatments (Lim et al., 1998; Rolli et al., 1999). Egr-1 has also been reported to be induced by administration of glutamate (Vaccarino et al., 1992), quinolinic acid (Massieu et al., 1992), Kaininc acid and anticonvulsant drugs such as pentylenetetrazol (Sonnenberg et al., 1989; Murphy et al., 1991), picrotoxin (Saffen et al., 1988), amphetamine (Cole et al., 1992), cocaine (Yamagata et al., 1994), lithium chloride (Lamprecht and Dudai, 1995) and formalin (Pertovaara et al., 1993).

The DNA binding activity of Egr-1 is altered in the presence of Zn, Fe, or Mn (Cao et al., 1990). Two forms of finger structures exist in Egr-1 proteins but only the metal-bound form can bind to its DNA site. The cellular redox state may control the DNA binding of Egr-1 by excluding Zn from the protein and rendering the oxidized form inactive (Huang and Adamson, 1993). Oxidation of Egr-1 excludes the metal from its complex. Metals such as iron, manganese, cadmium, magnesium, etc. can be incorporated in the finger structure replacing Zn (Cao et al., 1990).

Induction of zif268 by a PKC dependent mechanism was observed in PC12 cells after exposure to Pb (Kim et al., 2000). Exposure to Pb *in vivo* interferes with the regulatory mechanisms that control the expression of Egr-1, which may result in disruption of the program of gene expression that underlies neuronal development (Kim et al., 2002). Lactational exposure of rats to Pb resulted in modulation of Egr-1 DNA binding in different brain regions (Reddy and Zawia, 2000).

3.5 METAL-INDUCED ALTERATIONS OF GENERAL TRANSCRIPTION FACTORS

ZFP-driven transcriptional events are susceptible to environmental metals, and their disturbance may result in perturbations in the regulation of gene expression (Hanas and Gunn, 1996; Zawia et al., 1998, 2000; Reddy and Zawia, 2000). Hanas et al. (1999) have reported that the *in vitro* addition of 5–20 μ*M* Pb inhibited the DNA binding of TFIIIA to the 5S ribosomal gene, while 10–20 μ*M* Pb inhibited Sp1 DNA binding. Furthermore, DNase I protection assays by the same group showed that the inhibition by Pb was occurring at the zinc finger motif and that it was selective for zinc finger proteins. Cd, Pb, and arsenic exert inhibitory effects on the binding of the estrogen hormone receptor and TFIIIA (Simons et al., 1990; Predki et al., 1992; Hanas et al., 1999). Mercuric ions were also found to inhibit the DNA binding activity of TFIIIA and Sp1 (Rodgers et al., 2001).

The zinc finger structure is stabilized through binding of a Zn ion to two imidazole nitrogen (N) and two cysteine sulfhydryl (S) ligands. Since several metal ions exert high affinities towards SH- groups, potentially, zinc finger proteins could be targets of such ions. The repair inhibitions caused by Ni and Co were reversed by Mg, while Zn reversed the disturbance of DNA-protein interaction provoked by Cd (Asmuss et al., 2000). Each zinc finger protein appears to have its own structural features and sensitivities towards toxic metal ions.

3.6 CONCLUSIONS

Since zinc fingers are important and widespread, one would assume that environmental heavy metals should alter the transcription of every gene and result in widespread dysregulation of gene expression. However, we have found that the effects of Pb on gene expression are selective. Understanding the factors that impart selectivity and the conditions necessary for the deleterious effects of xenobiotic metals on zinc finger proteins is essential. Several conclusions may be inferred. There are codes that may impart selectivity to the action of xenobiotic metals on ZFP. One is the composition and arrangement of the Cys and His residues, the number of fingers in each protein, and the amino acids that constitute the immediate environment (sphere) of the finger. Another code may lie in the variety of GC box response elements that are recognized by the finger. The shape and composition of the finger will dictate the access and affinity of binding to xenobiotic metals. The DNA consensus sequence will determine recognition and DNA binding of such metallo-protein complexes. Thus, metals can interact with the zinc moiety but the consequences of such an interaction will depend on favorable dynamic equilibrium, access to the zinc site, and the ability of the cell to repair or counter any adverse effects.

Variation of responses and selectivity may also be a product of the interactions of specific heavy metals with selective signaling intermediates in the cell. Therefore, a heavy metal which interrupts a signaling pathway involved in the regulation of Sp1 DNA binding will impact the ability of Sp1 to activate gene expression. There are apparent differences among metals in their targeting of various signaling intermediates in the cell. Finally, a metal which has the ability to both directly act on the zinc finger domain and also interfere with signaling pathways that regulate Sp1 will have the greatest impact. Regardless of the mechanism involved, it is clear that exposure to heavy metals in the environment will disturb gene expression via actions on zinc finger proteins such as Sp1 and Egr-1.

ACKNOWLEDGMENTS

This research was supported by a grant from the National Institutes of Health (NIEHS, grant #ES08104).

REFERENCES

Ackerman, S.L. et al. (1991), Functional significance of an overlapping consensus binding motif for Sp1 and Zif268 in the murine adenosine deaminase gene promoter, *Proc. Natl. Acad. Sci. USA* 88, 7523–7527.

Ahmed, M.M. et al. (1997), Ionizing radiation-inducible apoptosis in the absence of p53 linked to transcription factor EGR-1, *J. Biol. Chem.* 272, 33056–33061.

Aschner, M. (1996), Astrocytes as modulators of mercury-induced neurotoxicity, *NeuroToxicology* 17, 663–669.

Asmuss, M., Mullenders, L.H., and Hartwig, A. (2000), Interference by toxic metal compounds with isolated zinc finger DNA repair proteins, *Toxicol. Lett.* 112–113, 227–231.

Bai, G., and Kusiak, J.W. (1995), Functional analysis of the proximal 5′-flanking region of the *N*-methyl-D-aspartate receptor subunit gene, NMDAR1, *J. Biol. Chem.* 270, 7737–7744.

Basha, M.R. et al. (2003), Lead-induced developmental perturbations in hippocampal Sp1 DNA binding are prevented by zinc supplementation: *in vivo* evidence for Pb and Zn competition, *Int. J. Dev. Neurosci.* 21, 1–12.

Beckmann, A.M., and Wilce, P.A. (1997), Egr transcription factors in the nervous system, *Neurochem. Int.* 31, 477–510.

Borellini, F., and Glazer, R.I. (1993), Induction of Sp1-p53 DNA binding heterocomplexes during granulocyte/macrophage colony-stimulating factor-dependent proliferation in human erythroleukemia cell line TF-1, *J. Biol. Chem.* 268, 7923–7928.

Briggs, M.R. et al. (1986), Purification and biochemical characterization of the promoter-specific transcription factor, Sp1, *Science* 234, 47–52.

Brydie, M. et al. (1999), Expression of the NMDA R1 subunit and its relationship to Sp1 DNA binding in the rat hippocampus following developmental lead exposure, *The Toxicologist* 53, 1156.

Bull, R. et al. (1983), The effects of lead on the developing central nervous system of the rat, *NeuroToxicology* 4, 1–18.

Busselberg, D. et al. (1991), Lead and zinc block a voltage-activated calcium channel of aplysia neurons, *J. Neurophysiol.* 65, 786–795.

Cao, X.M. et al. (1990), Identification and characterization of the Egr-1 gene product, a DNA binding zinc finger protein induced by differentiation and growth signals, *Mol. Cell. Biol.* 10, 1931–1939

Chavrier, P. et al. (1989), Structure, chromosome location, and expression of the mouse zinc finger gene Krox-20: multiple gene products and coregulation with the proto-oncogene c-fos, *Mol. Cell. Biol.* 9, 787–797.

Christy, B.A., Lau, L.F., and Nathans, D. (1988), A gene activated in mouse 3T3 cells by serum growth factors encodes a protein with "zinc finger" sequences, *Proc. Natl. Acad. Sci. USA* 85, 7857–7861.

Clarkson, T.W. (1997), The toxicology of mercury, *Crit. Rev. Clin. Lab. Sci.* 34, 369–403.

Cole, A.J. et al. (1992), D1 dopamine receptor activation of multiple transcription factor genes in rat striatum, *J. Neurochem.* 8, 1420–1426.

Crosby, S.D. et al. (1991), The early response gene NGFI-C encodes a zinc finger transcriptional activator and is a member of the GCGGGGGCG (GSG) element-binding protein family, *Mol. Cell. Biol.* 11, 3835–3841.

Crumpton, T. et al. (2001), Lead exposure in Pheochromocytoma (PC12) cells alters neural differentiation and Sp1 DNA binding, *NeuroToxicology* 22, 49–62.

DeFranco, C. et al. (1993), Nerve growth factor induces transcription of NGFIA through complex regulatory elements that are also sensitive to serum and phorbol 12-myristate 13-acetate, *Mol. Endocrinol.* 7, 365–379.

Desjarlais, J.R., and Berg, J.M. (1993), Use of a zinc-finger consensus sequence framework and specificity rules to design specific DNA binding proteins, *Proc. Natl. Acad. Sci. USA* 90, 2256–2260.

Dolle, P. et al. (1993), Local alteration of Krox-20 and Hox gene expression in the hindbrain suggest lack of rhombomeres 4 and 5 in homozygote null Hoxa-1 (Hox-1.6) mutant embryos, *Proc. Natl. Acad. Sci. USA* 90, 7666–7670.

Dynan, W.S., and Tjian, R. (1983a), Isolation of transcription factors that discriminate between different promoters recognized by RNA polymerase II, *Cell* 32, 669–680.

Dynan, W.S., and Tjian, R. (1983b), The promoter-specific transcription factor Sp1 binds to upstream sequences in the SV40 early promoter, *Cell* 35, 79–87.

Ebert, S.N., and Wong, D.L. (1995), Differential activation of the rat phenylethanolamine N-methyltransferase gene by Sp1 and Egr-1, *J. Biol. Chem.* 270, 17299–17305.

Gaiddon, C., Loeffler, J.P., and Larmet, Y. (1996), Brain-derived neurotrophic factor stimulates AP-1 and cyclic AMP-responsive element dependent transcriptional activity in central nervous system neurons, *J. Neurochem.* 66, 2279–2286.

Gencic, S., and Hudson, L.D. (1990), Conservative amino acid substitution in the myelin proteolipid protein of jimpymsd mice, *J. Neurosci.* 10, 117–124.

Gidoni, D, Dynan, W.S, and Tjian, R. (1984), Multiple specific contacts between a mammalian transcription factor and its cognate promoters, *Nature* 312, 409–413.

Gidoni, D. et al. (1985), Bidirectional SV40 transcription mediated by tandem Sp1 binding interactions, *Science* 230, 511–517.

Goyer, R.A. (1996), Toxic effects of metals. *Cassaret and Doull's Toxicology: The Science of Poisons*, McGraw-Hill, New York, 643–690.

Greene, L.A., and Shooter, E.M. (1980), The nerve growth factor: biochemistry, synthesis, and mechanism of action, *Annu. Rev. Neurosci.* 3, 353–402.

Guilarte, T.R., Miceli, R.C., and Jett, D.A. (1995), Biochemical evidence of an interaction of lead at the zinc allosteric sites of the NMDA receptor complex: effects of neuronal development, *NeuroToxicology* 16, 63–67.

Guroff, G. (1985), PC12 cells as a model of neuronal differentiation, *Cell Culture in the Neurosciences*, ed. J. Bottenstein and G. Sato, Plenum, New York, 1245–1272.

Hanas, J.S., and Gunn, C.G. (1996), Inhibition of transcription factor IIIA-DNA interactions by xenobiotic metal ions, *Nucleic Acids Res.* 24, 924–930.

Hanas, J.S. et al. (1999), Lead inhibition of DNA binding mechanism of Cys(2)His(2) zinc finger proteins, *Mol. Pharmacol.* 56, 982–988.

Harry, G.J. et al. (1996), Lead-induced alterations of glial fibrillary acidic protein (GFAP) in the developing rat brain, *Toxicol. Appl. Pharmacol.* 139, 84–93, 1996.

Hashemzadeh-Gargari, H., and Guilarte, T.R. (1999), Divalent cations modulate N-methyl-D-aspartate receptor function at the glycine site, *J. Pharmacol. Exp. Ther.* 290, 1356–1362.

Henson, J., Saffer, J., and Furneaux, H. (1992), The transcription factor Sp1 binds to the JC Virus promoter and is selectively expressed in glial cells in human brain, *Ann. Neurol.* 74, 72–78.

Huang, R.P., and Adamson, E.D. (1993), Characterization of the DNA binding properties of the early growth response-1 (Egr-1) transcription factor: evidence for modulation by a redox mechanism, *DNA Cell Biol.* 12, 265–273.

Joseph, L.J. et al. (1988), Molecular cloning, sequencing, and mapping of EGR2, a human early growth response gene encoding a protein with "zinc-binding finger" structure, *Proc. Natl. Acad. Sci. USA* 85, 7164–7168.

Kadonaga, J., and Tjian, R. (1986), Affinity purification of sequence-specific DNA binding proteins, *Proc. Natl. Acad. Sci. USA* 83, 5889–5893.

Kadonaga, J. et al. (1987), Isolation of cDNA encoding transcription factor SP1 and functional analysis of the DNA binding domain, *Cell* 51, 1079–1090.

Kern, M., and Audesirk, G. (1995), Inorganic lead may inhibit neurite development in cultured rat hippocampal neurons through hyperphosphorylation, *Toxicol. Appl. Pharmacol.* 134, 111–123.

Kim, K.A. et al. (2000), Immediate early gene expression in PC12 cells exposed to lead: requirement for protein kinase C, *J. Neurochem.* 74, 1140–1146.

Kim, K.A. et al. (2002), Exposure to lead elevates induction of zif268 and Arc mRNA in rats after electroconvulsive shock: the involvement of protein kinase C, *J. Neurosci. Res.* 69, 268–277.

Koizumi, S., and Otsuka, F. (1994), Nuclear proteins binding to the human metallothionein-IIA gene upstream sequences, *Ind. Health* 32, 193–205.

Komulainen, H., and Bondy, S.C. (1987), Increased free intrasynaptosomal Ca^{2+} by neurotoxic organmetals: distinctive mechanisms, *Toxicol. Appl. Pharmacol.* 88, 77–86.

Krizek, B.A., Zawadzke, L.E., and Berg, J.M. (1993), Independence of metal binding between tandem Cys2His2 zinc finger domains, *Protein Sci.* 2, 1313–1319.

Kumahara, E., Ebihara, T., and Saffen, D. (1999), Nerve growth factor induces zif268 gene expression via MAPK-dependent and -independent pathways in PC12D cells, *J. Biochem.* (Tokyo) 125, 541–553.

Lamprecht, R., and Dudai, Y. (1995), Differential modulation of brain immediate early genes by intraperitoneal LiCl, *Neuroreport* 7, 289–293.

Laterra, J. et al. (1992), Inhibition of astroglia-induced endothelial differentiation by inorganic lead: a role for protein kinase C, *Proc. Natl. Acad. Sci. USA* 89, 10748–10752.

Lee, M.S. et al. (1989), Three-Dimensional solution structure of a single Zinc-Finger DNA binding domain, *Science* 245, 635–637.

Lee, K.F. et al. (1992), Targeted mutation of the gene encoding the low affinity NGF receptor p75 leads to deficits in the peripheral sensory nervous system, *Cell* 69, 737–749.

Lemaire, P. et al. (1988), Two mouse genes encoding potential transcription factors with identical DNA binding domains are activated by growth factors in cultured cells, *Proc. Natl. Acad. Sci. USA* 85, 4691–4695.

Lim, R.W., Varnum, B.C., and Herschman, H.R. (1987), Cloning of tetradecanoyl phorbol ester-induced 'primary response' sequences and their expression in density-arrested Swiss 3T3 cells and a TPA non-proliferative variant, *Oncogene* 1, 263–270.

Lim, C.P., Jain, N., and Cao, X. (1998), Stress-induced immediate-early gene, egr-1, involves activation of p38/JNK1, *Oncogene* 16, 2915–2926.

Marin, M. et al. (1997), Transcription factor Sp1 is essential for early embryonic development but dispensable for cell growth and differentiation, *Cell* 89, 619–628.

Markovac, J., and Goldstein, G.W. (1988), Lead activates protein kinase C in immature rat brain microvessels, *Toxicol. Appl. Pharmacol.* 96, 14–23.

Massieu, L. et al. (1992), Administration of quinolinic acid in the rat hippocampus induces expression of c-fos and NGFI-A, *Mol. Brain Res.* 16, 88–96.

McCauley, P. et al. (1982), The effect of prenatal and postnatal lead exposure on neonatal synaptogenesis in the rat cerebral cortex, *J. Toxicol. Environ. Health* 10, 639–651.

Milbrandt, J. (1987), A nerve growth factor-induced gene encodes a possible transcriptional regulatory factor, *Science* 238, 797–799.

Mundy, W.R., Shafer, T.J., and Ward T.R. (1996), Pb^{2+} both activates and inhibits protein kinase C purified from rat brain and neuronal cell cultures. *Soc. Neurosci. Abstract.*

Mundy, W.R., Barone, S., and Parran, D.K. (2001), Effects of methylmercury and mercuric chloride on differentiation and cell viability in PC12 cells, *Toxicol. Sci.* 59, 278–290.

Murphy, T.H., Worley, P.F., and Baraban, J.M. (1991), L-type voltage-sensitive calcium channels mediate synaptic activation of immediate early genes, *Neuron* 7, 625–635.

Nagashima, K. (1997), Review of experimental methylmercury toxicity in rats: neuropathology and evidence for apoptosis, *Toxicol. Pathol.* 25, 624–631.

Nihei, M.K., and Guilarte, T.R. (1999), NMDAR-2A subunit protein expression is reduced in the hippocampus of rats exposed to Pb^{2+} during development, *Mol. Brain Res.* 66, 42–49.

Nichols, D.M., and McLachlan, D.R.C. (1990), Issues of Pb toxicity, *Ad. In Vivo Body Composition Studies*, ed. S. Yasumura, Plenum Press, New York, 237–246.

O'Conner, T.R. et al. (1993), Fpg protein of Escherichia Coli is a zinc finger protein whose cysteine residues have a structural and/or functional role, *J. Biol. Chem.* 268, 9063–9070.

O'Donovan, K.J. et al. (1999), The EGR family of transcription-regulatory factors: progress at the interface of molecular and systems neuroscience, *Trends Neurosci.* 22, 167–173.

Pabo, C.O., and Sauer, R.T. (1992), Transcription factors: structural families and principles of DNA recognition, *Annu. Rev. Biochem.* 61, 1053–1095.

Patwardhan, S. et al. (1991), EGR3, a novel member of the Egr family of genes encoding immediate-early transcription factors, *Oncogene* 6, 917–928.

Pavletich, N.P., and Pabo, C.O. (1991), Zinc finger-DNA recognition: crystal structure of a Zif268-DNA complex at 2.1 A, *Science* 252, 809–817.

Persengiev, S.P. et al. (1996), Transcription factor Sp1 expressed three different developmentally regulated mRNAs in mouse spermatogenic cells, *Endocrinology* 137, 638–646.

Pertovaara, A., Bravo, R., and Herdegen, T. (1993), Induction and suppression of immediate-early genes in the rat brain by a selective alpha-2-adrenoceptor agonist and antagonist following noxious peripheral stimulation, *Neuroscience* 54, 117–126.

Petit, T., Alfano, D., and LeBoutillier, J. (1983), Early lead exposure and the hippocampus: a review and recent advances, *NeuroToxicology* 4, 79–94.

Petit, T.L., and LeBoutillier, J.C. (1986), Zinc deficiency in the postnatal rat: implications for lead toxicity, *NeuroToxicology* 7, 237–246.

Predki, P.F., and Sarkar, B. (1992), Effect of replacement of "zinc finger" zinc on estrogen receptor DNA interactions, *J. Biol. Chem.* 267, 5842–5846.

Razmiafshari, M., and Zawia, N.H. (2000), Utilization of a synthetic peptide as a tool to study the interaction of heavy metals with the zinc finger domain of proteins critical for gene expression in the developing brain, *Toxicol. Appl. Pharmacol.* 166, 1–12.

Razmiafshari, M. et al. (2001), NMR identification of heavy metal-binding sites in a synthetic zinc finger peptide: toxicological implications for the interactions of xenobiotic metals with zinc finger proteins, *Toxicol. Appl. Pharmacol.* 172, 1–10.

Reddy, G.R., and Zawia, N.H. (2000), Lead exposure alters Egr-1 DNA binding in the neonatal rat brain, *Int. J. Dev. Neurosci.* 18, 791–795.

Rhodes, D., and Klug, A. (1993), Zinc finger structure, *Sci. Am.* 268, 32–39.

Rodgers, J.S. et al. (2001), Mercuric ion inhibition of eukaryotic transcription factor binding to DNA, *Biochem. Pharmacol.* 61, 1543–1550.

Rodier, P.M. (1986), Time of exposure and time of testing in developmental neurotoxicology, *NeuroToxicology* 7, 69–76.

Rodier, P. (1990), Critical periods for morphologic assessment, *Cong. Anom.* 32, 55–64.

Rolli, M. et al. (1999), Stress-induced stimulation of early growth response gene-1 by p38/stress-activated protein kinase 2 is mediated by a cAMP-responsive promoter element in a MAPKAP kinase 2-independent manner, *J. Biol. Chem.* 274, 19559–19564.

Saffen, D.W. et al. (1998), Convulsant-induced increase in transcription factor messenger RNAs in rat brain, *Proc. Natl. Acad. Sci. USA* 85, 795–799.

Saffer, J.D., Jackson, S.P., and Annarella, M.B. (1991), Developmental expression of Sp1 in the mouse, *Mol. Cell Biol.* 11, 2189–2199.

Sakamoto, K.M. et al. (1991), 5′ upstream sequence and genomic structure of the human primary response gene, EGR-1/TIS8, *Oncogene* 6, 867–871.

Shao, Z., and Suszkiw, J.B. (1991), Ca^{2+} surrogate action of Pb^{2+} on acetylcholine release from rat brain synaptosomes, *J. Neurochem.* 56, 568–574.

Silbergeld, E.K. (1992), Mechanisms of Pb neurotoxicity, or looking beyond the lamppost, *FASEB J.* 6, 3201–3206.

Simons, S.S. Jr., Chakraborti, P.K., and Cavanaugh, A.H. (1990), Arsenite and cadmium(II) as probes of glucocorticoid receptor structure and function, *J. Biol. Chem.* 265, 1938–1945.

Sonnenberg, J.L. et al. (1989), Glutamate receptor agonists increase the expression of Fos, Fra, and AP-1 DNA binding activity in the mammalian brain, *J. Neurosci. Res.* 24, 72–80.

Sunderman, F.W., Jr., and Barber, A.M. (1988), Finger-loops, oncogenes, and metals, *Ann. Clin. Lab. Sci.* 18, 267–288.

Suske, G. (1999), The Sp-family of transcription factors, *Gene* 238, 291–300.

Swiatek, P.J., and Gridley, T. (1993), Perinatal lethality and defects in hindbrain development in mice homozygous for a targeted mutation of the zinc finger gene Krox20, *Genes Dev.* 7(11), 2071–2084.

Thiel, G., Schoch, S., and Peterson, D. (1994), Regulation of synapsin-1 gene expression by the zinc finger transcription factor zif268/egr-1, *J. Biol. Chem.* 269, 15294–15301.

Thiel, G, Lietz, M, and Leichter, M. (1999), Regulation of neuronal gene expression, *Naturwissenschaften* 86, 1–7.

Thiesen, H.J., and Bach (1991), Transition metals modulate DNA-protein interactions of Sp1 zinc finger domains with its cognate target site, *Biochem. Biophys. Res. Commun.* 176, 551–557.

Tischler, A.S., and Greene, L.A. (1975), Nerve growth factor-induced process formation by cultured rat pheochromocytoma cells, *Nature* 258, 341–342.

Tomsig, J.L., and Suszkiw, J.B. (1996), Multisite interactions between Pb^{2+} and protein kinase C and its role in norepinephrine release from bovine adrenal chromaffin cells, *J. Neurochem.* 64, 2667–2673.

Tso, J.Y., Van Den Berg, D.J., and Korn, L.J. (1986), Structure of the gene for Xenopus transcription factor TFIIIA, *Nucleic Acids Res.* 14, 2187–2200.

Uvadia, A.J. et al. (1993), Sp-1 binds promoter elements regulated by the RB protein and Sp1-mediated transcription is stimulated by RB co-expression, *Proc. Natl. Acad. Sci. USA* 90, 3265–3268.

Vaccarino, F.M. et al. (1992), Differential induction of immediate early genes by excitatory amino acid receptor types in primary cultures of cortical and striatal neurons, *Brain Res. Mol. Brain Res.* 12, 233–241.

Verity, M.A. et al. (1994), Phospholipase A2 stimulation by methyl mercury in neuron culture, *J. Neurochem.* 62, 705–714.

Watson, M.A., and Milbrandt, J. (1990), Expression of the nerve growth factor-regulated NGFI-B genes in the developing rat, *Development* 110, 173–183.

Worley, P.F. et al. (1991), Constitutive expression of zif268 in neocortex is regulated by synaptic activity, *Proc. Natl. Acad. Sci. USA* 88, 5106–5110.

Yamagata, K. et al. (1994), Egr3/Pilot, a zinc finger transcription factor, is rapidly regulated by activity in brain neurons and colocalizes with Egr1/zif268, *Learn. Mem.* 1, 140–152.

Zawia, N.H., and Harry, G.J. (1995), Exposure to lead acetate modulates the developmental expression of myelin genes in the rat frontal lobe, *Int. J. Dev. Neurosci.* 13, 639–644.

Zawia, N.H., and Harry, G.J. (1996), Developmental exposure to Pb interferes with glial and neuronal differential gene expression in the rat cerebellum, *Toxicol. Appl. Pharmacol.* 138, 43–47.

Zawia, N.H. et al. (1998), SP1 as a target site for metal-induced perturbations of transcriptional regulation of developmental brain gene expression, *Brain Res.* 107, 291–298.

Zawia, N. H. et al. (2000), Disruption of the zinc finger domain: a common target that underlies many of the effects of lead, *NeuroToxicology* 21, 1–11.

CHAPTER 4

NF-κB and Neurotoxicity

C. A. Kassed, T. L. Butler, and Keith Pennypacker

CONTENTS

4.1 INTRODUCTION TO NF-κB

Much like in other physiological systems, homeostatic set points in the central nervous system (CNS) can shift in response to developmental cues, aging, the plasticity requirements of cognition and memory formation, or acute and chronic insults. Several well-characterized signaling pathways activate transcription in response to perturbations of homeostatic balance, but the Rel/Nuclear Factor-kappa B (NF-κB) family of transcription factors is distinct in its rapidity of activation and its unique mechanism of regulation. An organism's defense mechanisms require rapid signaling events that do not depend on *de novo* protein synthesis for coordination of gene expression to mount effective protective responses. Because an exceptionally large variety of stimuli and stressors requiring rapid reprogramming of gene expression can induce NF-κB within minutes, induction of this transcription factor is among the first lines of defense against threats to the health of an organism.

Disorders such as arthritis, cancer, chronic inflammation, asthma, heart disease, and some neuropathological conditions result from dysregulated NF-κB activity

(Gilmore et al., 1996; Foxwell et al., 1998; Makarov, 2000; Neve et al., 2000; Cechetto, 2001; Grabellus et al., 2002; Mou et al., 2002). NF-κB activity is also crucial to normal biological processes, including cellular proliferation, apoptosis, immune and inflammatory responses, and development, including that of the CNS (Muller et al., 1993; Schmidt-Ullrich et al., 1996; Bushdid et al., 1998; Lavrovsky et al., 2000; Lilienbaum et al., 2000; Mora et al., 2001; Toillon et al., 2002). An increased understanding of signal transduction pathways involved in normal and aberrant CNS functioning could assist in achieving the goal of enhancing recovery after specific types of neuropathological insult and conferring resistance to injury.

The term NF-κB commonly refers to the most avidly forming heterodimer of p50-RelA (p65) proteins, but the term applies to all dimeric forms of the Rel proteins. Identified in 1986 by Sen and Baltimore, NF-κB was originally detected as a B cell-specific transcription factor involved in controlling immunoglobulin (Ig) κ light chain gene expression, and was thus assumed to be tissue-restricted (Sen and Baltimore, 1986). NF-κB expression was later detected in other immune cells (Sen and Baltimore, 1986; Nabel and Baltimore, 1987; Pierce et al., 1988), and has subsequently been recognized as being ubiquitously expressed, yet cytoplasmically retained in a quiescent form until activated by appropriate signals (Siebenlist et al., 1994; Miyamoto and Verma, 1995; Baldwin, 1996; Ghosh et al., 1998). The notable importance of NF-κB in the immune system remains, with involvement in the regulation of mammalian genes encoding cytokines, cell adhesion molecules, anti-apoptotic factors, complement factors, and immunoreceptors. Table 4.1 contains a detailed list of NF-κB-regulated genes in many systems.

Rel proteins contain two highly conserved Ig-like domains in a region of ~300 amino acids known as the Rel homology region (RHR). There are five mammalian members of this family: NF-κB1 (p50 and its precursor 105), NFκB2 (p52 and its precursor p100), c-Rel, RelA (p65), and RelB (Siebenlist et al., 1994; Ghosh et al., 1998). The p105 and p100 precursor proteins possess both an RHR and a region with homology to the inhibitors of κB, or IκB proteins. The RHR has DNA recognition/binding function, dimerization, and nuclear localization functions, and is the region of interaction with IκB proteins (Bours et al., 1994; Schmitz et al., 1994; Baldwin, 1996). Rel family members can homo- or heterodimerize with other Rel proteins to form NF-κB dimers capable of DNA binding, but the various NF-κB dimers have different binding affinities for the κB sites contained in different genes (Urban and Baeuerle, 1991; Urban et al., 1991; Kunsch et al., 1992; Baeuerle and Henkel, 1994; Lin et al., 1995; Baldwin, 1996). In the consensus κB DNA binding sequence, 5'-GGGRNNYYCC-3', R is a purine, N is any base, and Y is a pyrimidine (Miyamoto and Verma, 1995).

Not all combinations of NF-κB subunits have been shown to form, and not all dimers that can form have physiological relevance. The major NF-κB complex in most cells is the p50-Rel A (p65) complex. DNA-binding sequence specificity is determined by the Rel dimer components, via the combination of different DNA recognition loops in the RHRs. As an example, although most abundant in most cell types, p50-p65 heterodimers cannot recognize the DNA binding motifs preferentially bound by p65 homodimers (Davis et al., 1991; Ganchi et al., 1993). Dimer composition is also thought to affect interactions with inhibitory/regulatory IκB proteins

Table 4.1 NF-κB-Regulated Genes

Class of Molecule	Target Gene
Cytokines/chemokines	Tumor necrosis factor (TNF)α,β
	IL-1, 2, 6, 8, 12
	β,γ-interferon
	RANTES
	MCP-1
	Gro -α,-β,-γ
Growth factors	Granulocyte/macrophage colony stimulating factor (G/M-CSF)
	Macrophage CSF
	Granulocyte CSF
Adhesion molecules	Mad-CAM-1
	ELAM-1
	ICAM-1
	VCAM
	E-Selectin
	P-Selectin
	C-reactive protein
Immunoreceptors	Ig kappa light chain
	Tissue factor-1
	β-2 Microglobulin
	T cell receptors
	Major histocompatibility complex class I,II
	T cell receptor β chain
	IL-2Rα
Viruses	HIV-1
	CMV
	SV-40
	Adenovirus
Acute phase proteins	Angiotensinogen
	Serum amyloid A precursor protein
	Complement factors (B, C4)
Transcriptional regulators	p53
	c-rel, v-rel
	IκBα
	NF-κB p105, p100, Bcl-3
	c-myc
Antiapoptotic	TRAF-1
	TRAF-2
	c-IAP1
	c-IAP2
	Bcl-xL
Enzymes	Nitric oxide synthase (NOS)
	Cyclooxygenase-2 (COX2)
	Phospholipase A2
	12-lipoxygenase
Other	A-20
	Vimentin
	Proteasomal LMP2
	TAP1

Table compiled from Lee and Burckart, 1998 and McKay and Cidlowski, 1999.

(Lenardo and Baltimore, 1989; Ganchi et al., 1993; Lernbecher et al., 1993; Baeuerle and Henkel, 1994; Siebenlist et al., 1994; Lin et al., 1995; Baldwin, 1996). This combinatorial diversity of Rel proteins, the ratio of each NF-κB subunit, and the temporal expression of NF-κB subunits all allow tight yet flexible gene regulation. Cell type specificity of the NF-κB response allows expression modulation of distinct, but overlapping sets of genes by altering expression patterns of Rel subunits.

The Rel proteins also differ in their capacity to activate transcription. NF-κB dimers must contain at least one Rel transactivation domain for *in vivo* transcriptional activation; proteins lacking transactivation domains, such as p50 homodimers, are believed to block NF-κB sites on DNA from transcriptionally active heterodimers (Muller and Harrison, 1995). NF-κB dimers that are transcriptionally active include RelB-p50, RelB-p52, p65-p50, p65-p65, and p65-c-Rel (Fujita et al., 1992; Ganchi et al., 1993; Grimm and Baeuerle, 1993; Moore et al., 1993; Sha et al., 1995).

All NF-κB complexes are tightly regulated through interactions with inhibitory IκB proteins. As with the Rel proteins, there are several IκB proteins with differing affinities for individual NF-κB dimers, as well as different cell type distributions. The mammalian family of 60–70 kDa IκB proteins to date includes IκBα, IκBβ, IκBγ, IκBε, Bcl-3, NF-κB1 (p105/p50), and NF-κB2 (p100/p52) (Whiteside and Israël, 1997; Ghosh et al., 1998). All IκBs contain six or seven ankyrin repeat motifs, novel stretches of 33 amino acids in tandem arrays of multiple copies, which are involved in protein–protein interaction between NF-κB and IκB. Importantly, only IκBα, IκBβ, and IκBε contain the NH_2-terminal regulatory regions required for stimulus-induced degradation, the essential step in activation of NF-κB. IκBα and IκBβ preferentially interact with dimers that contain p65, but are responsive to differing activating signals (Baldwin, 1996; Nakshatri et al., 1997). The p105 and p100 precursor proteins contain an NF-κB-like region and an IκB-like region. These molecules become shorter DNA binding proteins and IκB proteins by limited proteolysis or arrested translation (p105 to p50, and p100 to p52). The p50 and p52 become transcriptional modulators complexing with themselves or Rel proteins.

The best characterized IκB is IκBα, an inhibitor with gene expression regulated by NF-κB. The acidic region of IκBα blocks the DNA-binding/nuclear localization region of p65, resulting in an inactive, cytoplasmically sequestered NF-κB complex (Beg et al., 1992; Verma et al., 1995). IκBα also plays an important role in terminating NF-κB activation. The affinity of NF-κB for IκBα is higher than its affinity for κB sites on DNA, so when newly synthesized IκBα shuttles between cytoplasm and nucleus, NF-κB is bound and transported out of the nucleus via the strong nuclear export signal (NES) in IκB (Arenzana-Seisdedos et al., 1995; Tam et al., 2000). Regulation of IκBα is mediated by stimulus-induced phosphorylation of two NH_2-terminal serines (32 and 36). This is catalyzed by the macromolecular IκB kinase complex IKK and triggers rapid ubiquitination, followed by degradation of IκB by the 26S proteasome complex.

B cells contain constitutively active NF-κB, composed of p50-c-Rel heterodimers (Liou et al., 1994; Miyamoto et al., 1994). However, in most other mammalian cell types, activity of the ubiquitous, evolutionarily conserved NF-κB transcription factor is induced in response to a variety of stimuli (Baeuerle and Baltimore, 1996). The prototype p50–p65 NF-κB heterodimer is retained in the cytoplasm of nonstimulated cells complexed with an IκB inhibitor, most often IκBα (Siebenlist et al., 1994; Baldwin, 1996; Ghosh et al., 1998). IκB NH_2-terminal ankyrin repeats are positioned next to the NF-κB nuclear localization sequence (NLS) in the protein complex, sterically hindering NF-κB binding to nucleocytoplasmic karyopherin proteins that transport NF-κB to the nucleus (Huxford et al., 1998; Jacobs and Harrison, 1998; Pemberton et al., 1998). The NF-κB:IκB complex is disrupted within minutes in response to extracellular stimuli such as cytokines, T and B cell mitogens, viral proteins, double-stranded RNA, bacterial products, and chemical, physical, or oxidative stressors, through an energetically costly pathway (reviewed by Rothwarf and Karin, 1999). IκBα becomes phosphorylated at serines 32 and 36, then is targeted for ubiquitination at lysines 21 and 22 by a specialized E3 ubiquitin ligase (Traenckner et al., 1994; Brown et al., 1995; Verma et al., 1995; Krappmann et al., 1996; Yaron et al., 1998; Winston et al., 1999). Degradation of IκBα by the 26S proteasome follows, revealing the NLS of NF-κB and allowing its binding to karyopherin proteins (importins) (Morolanu, 1998). Freed, active NF-κB complexes translocate to the nucleus, bind cognate κB sequences of target genes, and modulate gene transcription.

Two other modes of NF-κB activation have been reported. Hypoxia induces phosphorylation of IκBα at Tyrosine 42, and the Src kinases have been implicated in this pathway of NF-κB activation (Imbert et al., 1996; Beraud et al., 1999). Subsequent dissociation of Tyr-phosphorylated IκBα from NF-κB may be mediated by interaction with phosphoinositide-3-kinase (PI3-kinase) rather than by a proteasome (Beraud et al., 1999). The other mode is exposure of cells to short wavelength ultraviolet radiation inducing IκBα degradation by the 26S proteasome, but this process does not involve serine or tyrosine phosphorylation (Bender et al., 1998; Li and Karin, 1998). NF-κB activation of these alternative pathways is much slower and weaker than the response to prototypical activators.

IKK is a complex comprised of two catalytic kinase subunits, the 85 kDa IKK1/α, and the 87 kDa IKKβ, plus the 48 kDa structural/regulatory subunit NEMO/IKKγ/IKKAP (Mercurio et al., 1999). Other proteins probably exist within the kinase, based on differences between the apparent mass of the purified complex and the total mass of the known subunits (Israël, 2000). Alternatively, it has been proposed that IKK is comprised of four catalytic subunits (Rothwarf and Karin, 1999). It has also been suggested that a number of different regulatory subunits exist that generate classes of IKK which respond to different upstream stimuli (Rothwarf and Karin, 1999; Karin and Delhase, 2000). The two IKK kinases are highly homologous, each containing an NH_2-terminal kinase domain, a leucine-zipper motif, and a helix-loop-helix (HLH) motif (Zandi et al., 1997). They form homo- or heterodimers and subsequently phosphorylate the two serine residues on IκB that target the inhibitor for degradation. Dimerization of IKK proteins is required for kinase activity of the complex, and each protein contains an activation loop with two serines that become phosphorylated in response to an extracellular signal (Delhase et al., 1999). Interestingly, serine phosphorylation of the activation loop in IKKβ, but not IKKα, is required for NF-κB activation by cytokines, suggesting a more important role for IKKβ (Mercurio et al., 1997). Most stimuli cause only transient IKK activation, with activity peaking within 5 to 15 minutes, then decreasing to ~25% of peak within 30 minutes (DiDonato et al., 1997; Zandi et al., 1997; Delhase et al., 1999). Autophosphorylation of IKKβ at a C-terminal serine cluster changes interaction between the HLH domain and the catalytic domain, resulting in decreased kinase activity and preventing prolonged activation of NF-κB (Delhase et al., 1999). Because a small reduction in IKK kinase activity can result in considerably decreased degradation of IκB, IKK is currently considered rate limiting for NF-κB activation (Rothwarf et al., 1998; Li et al., 1999; Rothwarf and Karin, 1999).

Efficient and effective defense by an organism necessitates maintenance of inducible NF-κB transcription, but also appropriate and timely termination or down-regulation of such transcription. IκBs inhibit DNA binding of NF-κB by retaining Rel complexes in the cytoplasm, so these proteins are the central regulators of NF-κB function. Because transcription of IκB is controlled by NF-κB, in effect this transcription factor can modulate its own expression. IκBs also regulate NF-κB in other ways. IκBα has both an unconventional NLS and a nuclear export sequence (NES) (Arenzana-Seisdedos et al., 1995; Sachdev and Hannink, 1998), and nucleus-cytoplasm shuttling of this inhibitor has been documented (Tam et al., 2000; Tam and Sen, 2001). IκBα seems to be involved in the removal of NF-κB from the nucleus (Zabel et al., 1993; Arenzana-Seisdedos et al., 1995), and this property of IκBα as an export chaperone appears to be required for NF-κB sequestration in the cytoplasm (Tam et al., 2000). IκBs can also be regulated by phosphorylation of residues in the COOH-terminal region, thereby limiting persistent NF-κB activation in most situations (Lin et al., 1996).

IKK is also central to down regulation of NF-κB action. Inactivation of IKK must occur to prevent transcription factor reactivation (via phosphorylation of IκB, then activation of NF-κB-induced transcription of IκB). Initial down-regulation occurs through COOH-terminal autophosphorylation of IKK, and continued decline of activity results from subsequent conformational changes in the COOH-terminal region (Delhase et al., 1999).

Chen et al. recently showed that RelA is subject to inducible acetylation, and that acetylated RelA interacts weakly with IκBα (Chen et al., 2001). Furia and others later demonstrated that p50 can also be acetylated, confirming that this post-translational modification plays a pivotal role in NF-κB regulation by controlling nuclear IκB binding to NF-κB (Fu et al., 2002; Furia et al., 2002). Deacetylation of RelA by histone deacetylase 3 (HDAC3) stimulates nuclear IκB binding to RelA, promoting termination of NF-κB activation followed by its export from the nucleus to the cytoplasm (Chen et al., 2001). The deacetylation-controlled response reestablishes latent cytoplasmic NF-κB:IκB complexes, preparing a cell for response to the next NF-κB-inducing stimulus.

4.2 NF-κB AND THE BRAIN

Cortical and hippocampal neurons contain low levels of constitutive NF-κB activity that are in concert with the high metabolic activity of these cells and the resulting endogenous antioxidants (Schütze et al., 1992; Kaltschmidt et al., 1993; Kaltschmidt et al., 1994). Activation and nuclear translocation of NF-κB occurs in the hippocampus following long-term potentiation (Meberg et al., 1996), during neuronal development (Schmidt-Ullrich et al., 1996; Guerrini et al., 1997), and as part of the injury response in neurons (Perez-Otaño et al., 1996). NF-κB activity is involved in the signaling required for maintenance of adult sensory neurons, as well as nerve regeneration (Wover et al., 2002). Baltimore et al. demonstrated activation of NF-κB in isolated synapses and nuclear translocation of p65 from dendrites to the nucleus. Also described was the differential localization of p50–p50 and p50–p65 dimers throughout the cytoplasm, with only p50–p65 localized at synapses (Baltimore et al., 2002). Several other groups have demonstrated involvement of NF-κB in signaling cascades at synaptic terminals (Duan et al., 1999; Albensi and Mattson, 2000; Freudenthal and Romano, 2000; Glazner et al., 2001).

Although it is tempting to assign a neuroprotective or pro-apoptotic role for activated NF-κB in the brain based upon evidence provided in a particular study, conclusive proof for a universal role for NF-κB signaling in the brain has been equivocal. When it is considered that a neuron receives simultaneous input from microglia, astrocytes, and dying and surviving neurons that can activate multiple signaling pathways, it becomes clear that consideration of only one of these contributions represents an oversimplified view. Increased neuronal NF-κB activity following brain injury can persist for at least 28 days (Penkowa et al., 2001; Kassed et al., 2002). Increases in NF-κB activity have also been reported in the context of some neuropathological conditions (Yan et al., 1996; Rostasy et al., 2000). Many studies have correlated brain-specific NF-κB upregulation with long-term changes in gene expression related to plasticity, repair, and expression of several neuroprotective genes, including manganese superoxide dismutase (MnSOD), calbindin, soluble forms of beta amyloid precursor protein (sAPP), the antiapoptotic protein Bcl-xL, and neurotrophins (Smith-Swintosky et al., 1994; Okazaki et al., 1995; Lahiri and Nall, 1995; Grimm et al., 1996; Meberg et al., 1996; Mattson et al., 2000; Cosgaya and Shooter, 2001; Glasgow et al., 2001; Hughes et al., 2001; Mattson and Camandola, 2001; Mattson and Chan, 2001). However, this pluripotent, multifunctional transcription factor has been linked to neuronal death in several models of cerebral ischemia (Salminen et al., 1995; Clemens, 2000).

Activation of NF-κB in neurons and in glia results from an extensive variety of stimuli. These include glutamate receptor agonists, depolarization, oxidative stress, bacterial lipopolysaccharide (LPS), double-stranded RNA, ceramide, neurotrophic factors, beta-amyloid peptide (Aβ), and cytokines (Cechetto, 2001). NF-κB has been associated with neuronal death in a number of neurodegenerative paradigms. Rapid increases in hippocampal NF-κB activity have been demonstrated within 4–16 hours after kainate-induced seizures in rats (Rong and Baudry, 1996). In a rodent cortical impact model of traumatic brain injury, NF-κB DNA binding activity was upregulated in the traumatized cortex for an extended period (Yang et al., 1995). Experiments by Terai revealed enhanced immunolabeling of NF-κB in glial cells of the penumbra in infarcted areas of post-mortem human brain (Terai et al., 1996). Nonaka observed NF-κB in microglia, macrophages, and astrocytes in the cortex of ischemic rats at 24 hours that was still detectable up to one year after injury (Nonaka et al., 1999). Clemens identified NF-κB expression in TUNEL- positive CA1 neurons 30 minutes after four vessel occlusion in rats (Clemens et al., 1997).

Conversely, experiments by Blondeau have indicated that rapidly increased NF-κB DNA-binding activity in neurons of rats after preconditioning with brief, sublethal ischemia, low-dose kainic acid, or linolenic acid conferred resistance against later episodes of ischemia or status epilepticus (Blondeau et al., 2001). *In vitro* experiments have shown that soluble forms of the beta-amyloid precursor protein (sAPP) activate NF-κB and rescue neurons from Aβ-induced death (Barger and Mattson, 1996a,b). A neuroprotective role for NF-κB against oxidative insults was also demonstrated by Lezoualc'h. In that study, an oxidative stress-resistant phenotype of PC12 cells was reversed by dexamethasone and a super-repressor mutant form of IκBα through suppression of NF-κB transcriptional activity, ultimately leading to increased cell death after H_2O_2 challenge (Lezoualc'h et al., 1998).

Altered expression of NF-κB has been demonstrated in the brains of patients with Alzheimer's disease (AD). Mao and Barger (1998) demonstrated that dehydroepiandrosterone (DHEA) declines during aging, causing decreases in NF-κB activity. DHEA declines were further exacerbated in patients with AD, correlating with lower NF-κB activity than in non-AD brain (Mao and Barger, 1998). On the other hand, studies utilizing immunohistochemical detection of NF-κB demonstrated markedly enhanced p65 staining in neurons, neurofibrillary tangles, dystrophic neurites, and some cells proximal to plaques in AD tissue of the hippocampus compared to non-AD, post-mortem brain (Terai et al., 1996; Kitamura et al., 1997). These studies were obviously not able to monitor changes in transcription factor activity, nor were expression levels of p50 measured, leaving the possibility that increases in the p65 subunit represented a compensation mechanism in response to pathological decreases in p50 activity or expression.

While it has been demonstrated that NF-κB activation can cause pro- and anti-apoptotic events, either outcome depends on the temporal sequence of signaling, the spatial/cell-type distribution, the individual gene promoters that are targeted, and the specific NF-κB proteins that are activated (Guo et al., 1998; Mattson et al., 1998; Lee et al., 1999; Lee et al., 1999; Yu et al., 1999; Yu et al., 2000; Duan et al., 2001). Varying survival times and diverse paradigms of experimental brain injury all likely contribute to conflicting evidence for the neurodegenerative versus the neuroprotective role of NF-κB, limiting the usefulness of broad conclusions concerning a universal NF-κB function. Different activating stimuli, variations in the timing and duration of activation, and diversity in the components of the DNA binding dimer likely prescribe NF-κB involvement in neuronal death or survival. Inappropriate activation or dysregulation of NF-κB almost certainly promotes neuronal degeneration after certain forms of brain injury or during the course of neuropathological diseases. An understanding of signal-transducing events in varying models of brain injury may further the possibility of pharmacologically modulating NF-κB to shift the balance of signals to neurons in favor of survival, thus enhancing structural plasticity and behavioral recovery after specific types of injury (Weber, 1999; Sharp et al., 2000).

4.3 THE TRIMETHYLTIN MODEL OF BRAIN INJURY

Damage control and adaptation to a post-injury environment in the CNS require activation of molecular and cellular events in an attempt to repair neuronal structures and re-establish connectivity through neurite sprouting. Insights into the molecular basis of selective resistance of specific neuronal populations to neurodegeneration have the potential to provide relief for neurotoxic brain injuries, perhaps even those caused by endogenous toxins. Intoxication by administration of trimethyltin (TMT) provides an excellent model for the analysis of these processes (Chang et al., 1983; Brock and O'Callaghan, 1987; Balaban et al., 1988). Because TMT is administered systemically, factors extrinsic to the brain such as peripheral immune cells are excluded as activators of the injury response, since there is no opportunity for frank infiltration through a compromised blood-brain barrier. This model of chemical-induced neurodegeneration provides an opportunity to assess the complex effects of a selective neurotoxicant on gene expression, the influence of genetic mutations on susceptibility to neurotoxicity, and the cellular mechanisms involved in injury resistance and neuronal repair.

Alkyltin compounds are intermediate byproducts in the production of plastics and preservatives, as well as ingredients in some miticides, algicides, bactericides, fungicides, and insecticides (WHO, 1980; Chang et al., 1982; Duménil, 1999). The organometallic compound TMT produces limbic-specific neurologic symptoms in humans and in experimental animals (Fortemps et al., 1978; Brown et al., 1979; Ross et al., 1981; Dyer et al., 1982; Bushnell and Evans, 1985; McMillan and Wenger, 1985). Although TMT can be formed by biomethylation of tin by estuaurine organisms, its CNS effects are currently more of interest as an experimental tool rather than a result of environmental toxicity. Despite the selectively with which TMT lesions the hippocampus, pyriform cortex, amygdaloid nucleus, and neocortex (Brown et al., 1979; Bouldin et al., 1981), not all limbic neurons are susceptible to TMT-induced damage (Krady et al., 1990; Toggas et al., 1992; Kassed et al., 2002), making it a useful paradigm for studying cellular processes associated with death of susceptible neurons and disturbed compartmentalization of neurotransmission (Andersson et al., 1997).

The set of symptoms produced by TMT intoxication has been referred to as "TMT syndrome." Manifestations of this toxicity include temporary anorexia and cessation of drinking with associated weight loss, hypoactivity followed by hyper-activity, hyperexcitability, somatosensory dysfunction, memory loss, impaired learning, and, at higher doses, ataxia, tremor, seizure, convulsion, and death (Dyer et al., 1982; Dey et al., 1997; Ekuta et al., 1998; Philbert et al., 2000). Responses to the toxic effects of TMT differ by strain as well as species and are well documented (Chang et al., 1983; Ekuta et al., 1998; Kassed et al., 2002), though the precise genetic basis for these effects is unknown. The dose-effect curve for TMT is significantly steeper for mice than for rats, and substantial differences exist between the two species in duration of sequelae, and the localization and timing of neuronal death (Chang et al., 1982; Chang and Dyer, 1983; Kassed et al., 2002). Pyramidal neurons of the rat CA3/CA4 and CA1 hippocampal subregions are sensitive to the effects of TMT, while in mice granule cells of the fascia dentata exhibit extensive damage (Chang, 1986; Bruccoleri et al., 1998; Fiedorowicz et al., 2001). In the mouse, some affected dentate granule neurons appear shrunken with pyknotic nuclei, while neuronal vacuolation can be observed in areas of severe neuronal loss (Chang et al., 1982; Kassed et al., 2002). Neurons that survive TMT-induced toxicity express

NF-κB, and many have eccentrically located nuclei with granulated cytoplasm (Chang et al., 1982; Kassed et al., 2002).

Neuronal damage caused by TMT is accompanied by microglial, then astroglial activation (McCann et al., 1996). There are some reports of cytokine expression after TMT, and these are somewhat controversial. A careful examination of the literature reveals that most studies demonstrating induction of cytokines by TMT were conducted *in vitro* using cultured glial cells (Maier et al., 1997; Viviani et al., 2001), or combinations of neurons and glia (Viviani et al., 1998) grown for differing periods of time prior to exposure. Animal sources for the glial cultures were 1- to 2-day-old rat pups, and neuronal cultures were obtained from fetal rats. The concentrations of TMT and exposure times varied widely in the *in vitro* experiments, and cytokine induction was measured only at early time points (6 hours or 24 hours), leaving the possibility that such induction might be transient. Responsiveness to TMT is influenced by the developmental stage of a cell; cultures containing astroblasts and neuronal precursors are very sensitive to even low concentrations of this toxicant, whereas differentiated cultures display a collection of responses with distinct, concentration-dependent changes in cell-type specific parameters (Monnet-Tschudi et al., 1995). The applicability to the intact adult animal of conclusions based upon exposure of cultured neurons and glia from immature animals to cytotoxic concentrations of TMT is tenuous.

Likewise for *in vivo* experiments, the influence of an animal's developmental stage on its response to toxic insult and induction of an immunological response cannot be ruled out. Studies using young mice assessed at early timepoints (Bruccoleri et al., 1998 (17 day old)); (Harry et al., 2000 (1 day)), or both (Fiedorowicz et al., 2001 (3 days after injection in 1-month-old mice)) have reported hippocampal cytokine induction after TMT administration. Increased susceptibility and heightened sensitivity to the effects of many toxicants in early youth and advanced age is common, documented, and often used in toxicological risk assessments for human populations (Irwin et al., 1992; Graeter and Mortenson, 1996 ; Savory et al., 1999; Howard and Pope, 2002). The younger and less mature the subject, the more different its response will be to toxicant exposure from that of an adult (Bruckner and Warren, 2001). Descriptions of the marked disparity in susceptibility to toxicity among rodents differing in age by just a few days date back to the 1960s (Done, 1964). Nevertheless, in adult mice seven days after a single injection of TMT, there was no evidence of proinflammatory cytokine induction (Kassed et al., 2002). Similar results were reported by Little et al. in adult rats at time points to 21 days (Little and O'Callaghan, 1999; Little et al., 2001; Little and O'Callaghan, 2001). These analyses demonstrate the usefulness of the TMT model of neuronal degeneration in adult rodents as one in which the confounding effects of inflammation and immune responses have no significant influence.

The granule cell damage induced by TMT in mice likely involves both necrosis and apoptosis (Harry et al., 1985; Chang, 1986; Bruccoleri et al., 1998; Bruccoleri et al., 1999; Fiedorowicz et al., 2001). This is an important aspect of the TMT model of degeneration, since only active cell death/apoptosis associated signaling pathways might be amenable to therapeutic intervention and are representative of neurodegenerative conditions such as AD, ischemic stroke, and excitotoxicity (Mattson et al.,

1998; Mattson et al., 2000). Cells in the CNS, spleen, and kidney that are sensitive to TMT contain the tin-binding protein stannin (Krady et al., 1990; Toggas et al., 1992; Dejneka et al., 1997). Stannin's normal intracellular function remains unclear, but decreased cell survival with increased caspase-3 activation was demonstrated in HeLa and PC-12 cells transfected with a stannin expression vector (Davidson et al., 2001). TMT is not metabolized, and blood levels of TMT peak within 1 hour of dosing, declining with a half-life of ~1.5 days (Ekuta et al., 1998). Blood to tissue transfer of TMT is relatively fast, with maximal distribution to the murine brain in 6 hours (Doctor et al., 1983). The low hemoglobin/tin binding capacity in mice (as in humans) accounts for a lack of TMT sequestration in murine blood, and is responsible for the exceptionally acute toxic profile compared to rats (Aldridge et al., 1981; Ekuta et al., 1998).

Damage to hippocampal neurons by TMT does not appear to be caused by a typical excitotoxic mechanism. Glutamate is released after exposure of depolarized hippocampal slices to TMT, and hippocampal glutamate levels decrease after a seizure-inducing injection of TMT to mice (Patel et al., 1990). However, blockade of NMDA receptors has no protective effect on TMT cytotoxicity (Andersson, 1996; Gunasekar et al., 2001). No significant changes are seen in the number of neuronal glutamate (NMDA) or kainate (KA) receptors or their binding 4 hours after TMT, but a delayed, extensive loss of receptors is observed 2 and 12 weeks after TMT. Levels of cytosolic calcium increase following TMT administration (Komulainen and Bondy, 1987), and may result from impaired activity of the Na, K-ATPase electrogenic pump (Vaccari et al., 1997; Stine et al., 1988; Gasso et al., 2000). Perturbations in serotonergic function and neurotransmission have been reported (Doctor et al., 1982; DeHaven et al., 1984; Andersson et al., 1995). The pattern of delayed excitatory disturbances is accompanied by spatially restricted increases in activity-dependent gene expression after TMT. These genes include brain-derived neurotrophic factor (BDNF), heat-shock protein 70 (hsp 70), and c-fos, and their pattern of induction by TMT differs from that typically seen after other forms of excitotoxic insults (Morgan et al., 1987; Le Gal La Salle, 1988). The TMT-induced increase in BDNF mRNA is suggestive of hyperactivity, but BDNF can be regulated by non-NMDA receptors, and stimulation of the perforant path may be the cause of such increases (Patterson et al., 1992). Progressive, TMT-induced mitochondrial dysfunction could result in reduced ATP production and impaired regulation of excitatory neurotransmission, as suggested by Albin (Albin and Greenamyre, 1992). Alternatively, neurons may require increased Na, K-ATPase activity as a survival response of neurons after TMT, leading to a reduction of ATP (Kassed et al., 2002).

Decreased mRNA expression for some GABAα and GABAβ receptor subunits parallels the pattern of neuronal degeneration in the CA3 subregion after TMT, while expression of the message for GABAα4 is increased (Nishimura et al., 2001; Nishimura et al., 2001). Pathological alterations in GABA-ergic basket cells and hilar interneurons are observable after TMT (Chang and Dyer, 1985), yet the number of GABA-immunoreactive neurons does not change (Andersson et al., 1994). Impairments in inhibitory neurotransmission appear to be involved in TMT toxicity, but are likely not the sole cause of damage. Reduced recurrent inhibition of dentate granule cells has been detected after TMT (Dyer, 1982; Dyer and Boyes, 1984), but enhancement of GABA-ergic function does not affect neurotoxicity. 5-HT and L-DOPA levels are not different in TMT-treated rats relative to controls, indicating that monoamine synthesis is not affected (Andersson et al., 1995).

The susceptibility of neurons to oxidative damage has prompted several groups to investigate the formation of reactive oxygen species (ROS) in the hippocampus after TMT treatment (Ali et al., 1992; Gunasekar et al., 2001; Viviani et al., 2001). Antioxidants can partially reverse the TMT lesion (Andersson, 1996), while catalase and the nitric oxide synthase inhibitor N^G-nitro-L-arginine methyl ester (L-NAME) can inhibit generation of hydrogen peroxide (H_2O_2) and nitric oxide (NO) by TMT (Gunasekar et al., 2001). Viviani has shown that TMT-induced synthesis of prostaglandin E_2 was abolished by indomethacin (Viviani et al., 2001), implicating activation of cyclooxygenase by TMT. It has been well established that activation of the arachidonic acid cascade is always associated with production of ROS (Van Kessel et al., 1987; Muller and Sorrell, 1997). Therefore, although the mechanism of TMT-induced neuronal injury in the hippocampus remains unclear, oxidative events probably play a considerable role. TMT reliably reproduces distinct alterations associated with pathological CNS lesions that are both characteristic of neurodegenerative diseases and reminiscent of the inefficient functioning of maintenance processes crucial for neuronal survival and plasticity associated with aging, including oxidative stress.

4.4 TMT AND NF-κB

NF-κB signal transduction has been studied in TMT-induced hippocampal neurodegeneration. The function of the NF-κB p50 subunit has been debated as a neurosurvival or neurodegenerative signal in the hippocampus. Phosphorylation of IκBα is present at 1 day after TMT administration, prior to activation of the p50 subunit in rat hippocampal neurons at 2 days after TMT treatment (Pennypacker et al., 2001). Moreover, activated p50, phosphorylated IκBα and IκB kinases reside in the cytoplasm and processes of hippocampal neurons demonstrating colocalization of NF-κB signaling molecules. Activated p50 is not colocalized with neurodegenerative markers, strongly suggesting that it is expressed in surviving neurons (Kassed et al., 2002).

To conclusively show that p50 serves a neuroprotective function in the hippocampus, mice lacking expression of the p50 protein were treated with TMT (Kassed et al., 2002). Lack of p50 increased the number of degenerating hippocampal neurons fivefold over the non-transgenic control mice. Furthermore, p50 becomes activated prior to and remains activated long after neurodegeneration has ceased. Thus, p50 activation functions as a neuroprotective signal in the TMT model of hippocampal neurodegeneration.

4.5 SUMMARY

Neurons are architecturally complex, containing extensive subcellular compartmentalization essential for the exquisite specificity of signal transduction in the brain.

During the injury process, a neuron receives input from both survival and death signals, and the balance of these opposing signals may determine the fate of the affected neurons (Xia et al., 1995). Integration of pro- and anti-apoptotic signals is pivotal to neuronal fate, and elucidating these pathways may eventually permit pharmacological shifting of homeostasis to favor neuronal survival after some pathological insults or pro-degenerative conditions. The p50 subunit of NF-κB is a signal in the hippocampus that tips the balance in favor of neurosurvival after TMT-induced injury. Understanding the molecular activation of this transcription factor will elucidate signaling proteins to be targeted for pharmacological intervention to favor neurosurvival after a neurotoxic insult or other types of injury to the brain.

ACKNOWLEDGMENTS

This study was supported by NIH Grant RO1 NS39141-01A2, American Heart Association Grant 9930072N (KRP), and American Heart Association Grant 0120233B (TLB).

REFERENCES

Albensi, B., and Mattson, M. (2000), Evidence for involvement of TNF and NF-kappaB in hippocampal and synaptic plasticity, *Synapse* 35, 151–159.

Albin, R., and Greenamyre, J. (1992), An alternative excitotoxic hypothesis, *Neurology* 42, 733–738.

Aldridge, W. N. et al. (1981), Brain damage due to trimethyltin compounds, *Lancet* 2, 692–693.

Ali, S., Lebel, C., and Bondy, S. (1992), Reactive oxygen species formation as a biomarker of methylmercury and trimethyltin in neurotoxicology, *Neurotoxicology* 13, 637–648.

Andersson, H. et al. (1995), Time-course of trimethyltin effects on the monoaminergic systems of the rat brain. *NeuroToxicology* 16, 201–210.

Andersson, H., Luthman, J., and Olson, L. (1994), Trimethyltin-induced expression of GABA and vimentin immunoreactivities in astrocytes of the rat brain, *Glia* 11, 378–82.

Andersson, H. et al. (1997), Trimethyltin exposure in the rat induces delayed changes in brain-derived neurotrophic factor, fos and heat shock protein 70, *NeuroToxicology* 18, 147–159.

Arenzana-Seisdedos, F. et al. (1995), Inducible nuclear expression of newly synthesized IkBa negatively regulates DNA-binding and transcriptional activities of NF-kappa B, *Mol. Cell Biol.* 15, 2689–2696.

Baeuerle, P., and Henkel, T. (1994), Function and activation of NF-KB in the immune system, *Annu. Rev. Immunol.* 12, 141–179.

Baeuerle, P. A., and Baltimore, D. (1996), NF-kappa B: ten years after, *Cell* 87, 13–20.

Balaban, C. D., O'Callaghan, J. P., and Billingsley, M. L. (1988), Trimethyltin-induced neuronal damage in the rat brain: comparative studies using silver degeneration stains, immunocytochemistry and immunoassay for neuronotypic and gliotypic proteins, *Neuroscience* 26, 337–361.

Baldwin, A. (1996), The NF-kappaB and I kappa B proteins: new discoveries and insights, *Annu. Rev. Immunol.* 14, 649–683.

Barger, S., and Mattson, M. (1996a), Induction of neuroprotective kappa B-dependent transcription by secreted forms of the Alzheimer's beta-amyloid precursor, *Brain Res.
Mol. Brain Res.* 40, 116–126.

Barger, S., and Mattson, M. (1996b), Participation of gene expression in the protection against
amyloid beta-peptide toxicity by the beta-amyloid precursor protein, *Ann. N.Y. Acad.
Sci.* 777, 303–309.

Beg, A. et al. (1992), IkB interacts with the nulcear localization sequences of the subunits of
NF-kB: a mechanism for cytoplasmic retention, *Genes Dev.* 6, 1899–1913 (with
erratum in *Genes Dev.* [1992] 6, 2664–2665).

Bender, K. et al. (1998), Sequential DNA damage-independent and -dependent activation of
NF-kappa B by UV, *EMBO J.* 17, 5170–5180.

Beraud, C., Henzel, W. J., and Baeuerle, P. A. (1999), Involvement of regulatory and catalytic
subunits of phosphoinositide 3-kinase in NF-kappaB activation, *Proc. Natl. Acad.
Sci. USA* 96, 429–34.

Blondeau, N. et al. (2001), Activation of the nuclear factor-kappaB is a key event in brain
tolerance, *J. Neurosci.* 21, 4668–4677.

Bouldin, T. et al. (1981), Pathogenesis of trimethyltin neuronal toxicity, *Am. J. Pathol.* 104,
237–249.

Bours, V. et al. (1994), Human RelB (I-Rel) functions as a kB site-dependent transactivating
memeber of the family of Rel-related proteins, *Oncogene* 9, 1699–1702.

Brock, T. O., and O'Callaghan, J. P. (1987), Quantitative changes in the synaptic vesicle
proteins synapsin I and p38 and the astrocyte-specific protein glial fibrillary acidic
protein are associated with chemical-induced injury to the rat central nervous system,
J. Neurosci. 7, 931–942.

Brown, A. et al. (1979), The behavioral and neuropathologic sequelae of intoxication by
trimethyltin compounds in the rat, *Am. J. Pathol.* 97, 61–76.

Brown, K. et al. (1995), Control of IkBa proteolysis by site-specific signal-induced phosphorylation, *Science* 267, 1485–1488.

Bruccoleri, A., Brown, H., and Harry, G. (1998), Cellular localization and temporal elevation
of tumor necrosis factor-a, interleukin-1a, and transforming growth factor-b1 mRNA
in hippocampal injury response induced by trimethyltin, *J. Neurochem.* 71,
1577–1587.

Bruccoleri, A., Pennypacker, K. R., and Harry, G. J. (1999), Effect of dexamethasone on
elevated cytokine mRNA levels in chemical-induced hippocampal injury, *J. Neurosci.
Res.* 57, 916–926.

Bruckner, J., and Warren, D. (2001), Toxic effects of solvents and vapors, *Casaret and Doul's
Toxicology, The Basic Science of Poisons*, ed. C. Klassen, McGraw-Hill, New York,
869–916.

Bushdid, P. B. et al. (1998), Inhibition of NF-kappaB activity results in disruption of the
apical ectodermal ridge and aberrant limb morphogenesis, *Nature* 392, 615–618.

Bushnell, P. J., and Evans, H. L. (1985), Effects of trimethyltin n homecage behavior of rats,
Toxicol. Appl. Pharmacol. 79, 134–142.

Cechetto, D. (2001), Role of nuclear factor kappa B in neuropathological mechanisms, *Prog.
Brain Res.* 132, 391–404.

Chang, L., and Dyer, R. (1983), A time-course study of trimethyltin induced neuropathology
in rats, *Neurobehav. Toxicol. Teratol.* 5, 443–459.

Chang, L. et al. (1982), Neuropathology of mouse hippocampus in acute trimethyltin intoxication, *Neurobehav. Toxicol. Teratol.* 4, 149–156.

Chang, L. et al. (1982), Neuropathology of trimethyltin intoxication. I. Light microscopy
study, *Environ. Res.* 29, 435–444.

Chang, L. et al. (1982), Neuropathology of trimethyltin intoxication. II. Electron microscopic study on the hippocampus, *Environ. Res.* 29, 445–458.

Chang, L. W. (1986), Neuropathology of trimethyltin: a proposed pathogenetic mechanism, *Fundam. Appl. Toxicol.* 6, 217–232.

Chang, L. W., and Dyer, R. S. (1985), Early effects of trimethyltin on the dentate gyrus basket cells: a morphological study, *J. Toxicol. Environ. Health* 16, 641–653.

Chang, L. W. et al. (1983), Species and strain comparison of acute neurotoxic effects of trimethyltin in mice and rats, *Neurobehav. Toxicol. Teratol.* 5, 337–350.

Chen, L.-f. et al. (2001), Duration of Nuclear NF-kB action regulated by reversible acetylation, *Science* 293, 1653–1657.

Clemens, J. A. (2000), Cerebral ischemia: gene activation, neuronal injury, and the protective role of antioxidants, *Free Radic. Biol. Med.* 28, 1526–1531.

Clemens, J. A. et al. (1997), Global cerebral ischemia activates nuclear factor-kappa B prior to evidence of DNA fragmentation, *Brain Res. Mol. Brain Res.* 48, 187–196.

Cosgaya, J. M., and Shooter, E. M. (2001), Binding of nerve growth factor to its p75 receptor in stressed cells induces selective IkappaB-beta degradation and NF-kappaB nuclear translocation, *J. Neurochem.* 79, 391–319.

Davidson, C., Conn, R., and Billingsley, M. (2001), Evidence that stannin mediates organometal toxicity through activation of caspase-3, *Soc. Neurosci. Abs.* 874.13.

Davis, N. et al. (1991), Rel-associated pp40: an inhibitor of the Rel family of transcription factors, *Science* 253, 1268–1271.

DeHaven, D., Walsh, T., and Mailman, R. (1984), Effects of trimethyltin on dopaminergic and serotonergic function in the central nervous system, *Toxicol. Appl. Pharmacol.* 74, 182–189.

Dejneka, N. S. et al. (1997), Localization and characterization of stannin: relationship to cellular sensitivity to organotin compounds, *Neurochem. Int.* 31, 801–815.

Delhase, M. et al. (1999), Positive and negative regulation of IkB kinase activity through IKKb subunit phosphorylation, *Science* 284, 309–313.

Dey, P. et al. (1997), Altered expression of polysialated NCAM in mouse hippocampus following trimethyltin administration, *NeuroToxicology* 18, 633–644.

DiDonato, J. et al. (1997), A cytokine-responsive IkB kinase that activates transcription factor NF-kB, *Nature* 388, 548–554.

Doctor, S. V. et al. (1982), Trimethyltin inhibits uptake of neurotransmitters into mouse forebrain synaptosomes, *Toxicology* 25, 213–21.

Doctor, S. V., Sultatos, L. G., and Murphy, S. D. (1983), Distribution of trimethyltin in various tissues of the male mouse, *Toxicol. Lett.* 17, 43–48.

Done, A. (1964), Developmental pharmacology. *Clin. Pharmacol. Ther.* 5, 432–479.

Duan, W., Guo, Z., and Mattson, M. P. (2001), Brain-derived neurotrophic factor mediates an excitoprotective effect of dietary restriction in mice, *J. Neurochem.* 76, 619–626.

Duan, W., Rangnekar, V., and Mattson, M. (1999), Prostate apoptosis response-4 production in synaptic compartments following apoptotic and excitotoxic insults: evidence for a pivotal role in mitochondrial dysfunction and neuronal degeneration, *J. Neurochem.* 72, 2312–2322.

Dyer, R. et al. (1982), The trimethyltin syndrome in rats, *Neurobehav. Toxicol. Teratol.* 4, 127–133.

Dyer, R. S. (1982), Physiological methods for assessment of Trimethyltin exposure, *Neurobehav. Toxicol. Teratol.* 4, 659–664.

Dyer, R. S., and Boyes, W. K. (1984), Trimethyltin reduces recurrent inhibition in rats, *Neurobehav. Toxicol. Teratol.* 6, 367–71.

Ekuta, J. E., Hikal, A. H., and Matthews, J. C. (1998), Toxicokinetics of trimethyltin in four inbred strains of mice, *Toxicol. Lett.* 95, 41–46.

Fiedorowicz, A. et al. (2001), Dentate granule neuron apoptosis and glia activation in murine hippocampus induced by trimethyltin exposure, *Brain Res.* 912, 116–127.

Fortemps, E. et al. (1978), Trimethyltin poisoning. Report of two cases, *Int. Arch. Occup. Environ. Health* 41, 1–6.

Foxwell, B. et al. (1998), Efficient adenoviral infection with IkappaB alpha reveals that macrophage tumor necrosis factor alpha production in rheumatoid arthritis is NF-kappaB dependent, *Proc. Natl. Acad. Sci. USA* 95, 8211–8215.

Freudenthal, R., and Romano, A. (2000), Participation of Rel/NF-kappaB transcription factors in long-term memory in the crab Chasmagnathus, *Brain Res.* 855, 274–281.

Fu, M. et al. (2002), Androgen receptor acetylation governs *trans* activation and MEKK1-induced apoptosis without affecting *in vitro* sumoylation and *trans*-repression function, *Mol. Cell. Biol.* 22, 3373–3388.

Fujita, T. et al. (1992), Independent modes of transcriptional activation by the p50 and p65 subunits of NF-kB, *Genes Dev.* 6, 775–787.

Furia, B. et al. (2002), Enhancement of nuclear factor-kappa B acetylation by coactivator p300 and HIV-1 Tat proteins, *J. Biol. Chem.* 277, 4973–4980.

Ganchi, P. et al. (1993), A novel NF-kB complex containing p65 homodimers: implications for transcriptional control at the level of subunit dimerization, *Mol. Cell Biol.* 13, 7826–7835.

Gasso, S. et al. (2000), Trimethyltin and triethyltin differentially induce spontaneous noradrenaline release from rat hippocampal slices, *Toxicol. Appl. Pharmacol.* 62, 189–196.

Ghosh, S., May, M., and Kopp, E. (1998), NF-kB and Rel proteins: evolutionarily conserved mediators of immune responses, *Annu. Rev. Immunol.* 16, 225–260.

Gilmore, T. et al. (1996), Rel/NF-kappaB/I kappaB proteins and cancer, *Oncogene* 13, 1367–1378.

Glasgow, J. N. et al. (2001), Transcriptional regulation of the BCL-X gene by NF-kappaB is an element of hypoxic responses in the rat brain, *Neurochem. Res.* 26, 647–659.

Glazner, G. W. et al. (2001), Endoplasmic reticulum D-myo-inositol 1,4,5-trisphosphate-sensitive stores regulate nuclear factor-kappaB binding activity in a calcium-independent manner, *J. Biol. Chem.* 276, 22461–22467.

Grabellus, F. et al. (2002), Reversible activation of nuclear factor-kappaB in human end-stage heart failure after left ventricular mechanical support, *Cardiovasc. Res.* 53, 124–130.

Graeter, L., and Mortenson, M. (1996), Kids are different: developmental variability in toxicology, *Toxicology* 111, 15–20.

Grimm, S., and Baeuerle, P. A. (1993), The inducible transcription factor NF-kappa B: structure-function relationship of its protein subunits, *Biochem. J.* 290, 297–308.

Grimm, S. et al. (1996), Bcl-2 down-regulates the activity of transcription factor NF-kappaB induced upon apoptosis, *J. Cell Biol.* 134, 13–23.

Guerrini, L. et al. (1997), Glutamate-dependent activation of NF-kappaB during mouse cerebellum development, *J. Neurosci.* 17, 6057–6063.

Gunasekar, P. et al. (2001), Mechanisms of the apoptotic and necrotic actions of trimethyltin in cerebellar granule cells, *Toxicol. Sci.* 64, 83–89.

Gunasekar, P. G. et al. (2001), Role of astrocytes in trimethyltin neurotoxicity, *J. Biochem. Mol. Toxicol.* 15, 256–262.

Guo, Q., Robinson, N., and Mattson, M. P. (1998), Secreted beta-amyloid precursor protein counteracts the proapoptotic action of mutant presenilin-1 by activation of NF-kappaB and stabilization of calcium homeostasis, *J. Biol. Chem.* 273, 12341–12351.

Harry, G. J. et al. (1985), The use of Synapsin I as a biochemical marker for neuronal damage by trimethyltin, *Brain Res.* 326, 9–18.

Harry, G. J. et al. (2000), Age-dependent cytokine responses: trimethyltin hippocampal injury in wild-type, APOE knockout, and APOE4 mice, *Brain Behav. Immun.* 14, 288–304.

Howard, M. D., and Pope, C. N. (2002), In vitro effects of chlorpyrifos, parathion, methyl parathion and their oxons on cardiac muscarinic receptor binding in neonatal and adult rats, *Toxicology* 170, 1–10.

Hughes, A. L. et al. (2001), Distinction between differentiation, cell cycle, and apoptosis signals in PC12 cells by the nerve growth factor mutant delta9/13, which is selective for the p75 neurotrophin receptor, *J. Neurosci. Res.* 63, 10–19.

Huxford, T. et al. (1998), The crystal structure of the IkBa/NF-kB complex reveals mechanisms of NF-kB inactivation, *Cell* 95, 759–770.

Imbert, V. et al. (1996), Tyrosine phosphorylation of I kappa B-alpha activates NF-kappa B without proteolytic degradation of I kappa B-alpha, *Cell* 86, 787–798.

Irwin, I. et al. (1992), The relationships between aging, monoamine oxidase, striatal dopamine and the effects of MPTP in C57BL/6 mice: a critical reassessment, *Brain Res.* 572, 224–231.

Israël, A. (2000), The IKK complex: an integrator of all signals that activate NF-kB? *Trends Cell Biol.* 10, 129–133.

Jacobs, M., and Harrison, S. (1998), Structure of an IkBa/NF-kB complex, *Cell* 95, 749–758.

Kaltschmidt, C., Kaltschmidt, B., and Baeuerle, P. A. (1993), Brain synapses contain inducible forms of the transcription factor NF- kappa B, *Mech. Dev.* 43, 135–147.

Kaltschmidt, C. et al. (1994), Constitutive NF-kappa B activity in neurons, *Mol. Cell Biol.* 14, 3981–3992.

Karin, M., and Delhase, M. (2000), The IkB kinase (IKK) and NF-kB: key elements of proinflammatory signalling, *Semin. Immunol.* 12, 85–98.

Kassed, C. et al. (2002), Lack of NF-kB p50 exacerbates degeneration of hippocampal neurons after chemical exposure and impairs learning, *Exp. Neurol.* 176, 277–288.

Kitamura, Y. et al. (1997), Alteration of transcription factors NF-kappaB and STAT1 in Alzheimer's disease brains, *Neurosci. Lett.* 237, 17–20.

Komulainen, H., and Bondy, S. (1987), Increased free intrasynaptosomal Ca+2 by neurotoxic organometals: distinctive mechanisms, *Toxicol. Appl. Pharmacol.* 88, 77–86.

Krady, J. et al. (1990), Use of avidin-biotin subtractive hybridization to characterize mRNA common to neurons destroyed by the selective neurotoxicant trimethyltin, *Mol. Brain Res.* 7, 287–297.

Krappmann, D., Wulczyn, F., and Scheidereit, C. (1996), Different mechanisms control signal-induced degradation and basal turnover of the NF-kB inhibitor IkBa *in vivo*, *EMBO J.* 15, 6716–6726.

Kunsch, C., Ruben, S., and Rosen, C. (1992), Selection of optimal kB Rel DNA-binding motifs: interaction of both subunits of NF-kB with DNA is required for transcriptional activation, *Mol. Cell Biol.* 12, 4412–4421.

Lahiri, D., and Nall, C. (1995), Promoter activity of the gene encoding the beta-amyloid precursor protein is up-regulated by growth factors, phorbol ester, retinoic acid and interleukin-1, *Brain Res. Mol. Brain Res.* 32, 233–240.

Lavrovsky, Y. et al. (2000), Role of redox-regulated transcription factors in inflammation, aging and age-related diseases, *Exp. Gerontol.* 35, 521–532.

Le, Gal La Salle G. (1988), Long-lasting and sequential increase of c-fos oncoprotein expression in kainic acid-induced status epilepticus, *Neurosci. Lett.* 88, 127–130.

Lee, J. I., and Burckart, G. J. (1998), Nuclear factor kappa B: important transcription factor and therapeutic target, *J. Clin. Pharmacol.* 38, 981–993.

Lee, H. H. et al. (1999), Specificities of CD40 signaling: involvement of TRAF2 in CD40-induced NF-kappaB activation and intercellular adhesion molecule-1 up-regulation, *Proc. Natl. Acad. Sci. USA* 96, 1421–1426.

Lee, S. J. et al. (1999), Transcriptional regulation of the intercellular adhesion molecule-1 gene by proinflammatory cytokines in human astrocytes, *Glia* 25, 21–32.

Lenardo, M., and Baltimore, D. (1989), NF-kB: a pleiotropic mediator of inducible and tissue-specific gene control, *Cell* 58, 227–229.

Lernbecher, T., Muller, U., and Wirth, T. (1993), Distinct NF-kB/Rel transcription factors are responsible for tissue-specific and inducible gene activation. Similarity to TA1 and phorbol ester-stimulated activity and phosphorylation in intact cells, *Nature* 365, 767–770.

Lezoualc'h F. et al. (1998), High constitutive NF-kappa B activity mediates resistance to oxidative stress in neuronal cells, *J. Neurosci.* 18, 3224–3232.

Li, N., and Karin, M. (1998), Ionizing radiation and short wave length UV activate NF-kB through two distinct mechanisms, *Proc. Natl. Acad. Sci. USA* 95, 13012–13017.

Li, Z. W. et al. (1999), The IKKbeta subunit of IkappaB kinase (IKK) is essential for nuclear factor kappaB activation and prevention of apoptosis, *J. Exp. Med.* 189, 1839–1845.

Lilienbaum, A. et al. (2000), NF-kappa B is developmentally regulated during spermatogenesis in mice, *Dev. Dyn.* 219, 333–340.

Lin, R., Gewert, D., and Hiscott, J. (1995), Differential transcriptional activation *in vitro* by NF-kB/Rel proteins, *J. Biol. Chem.* 270, 3123–3131.

Lin, R. et al. (1996), Phosphorylation of IkappaB in the C-terminal PEST domain by casein kinase II affects intrinsic protein stability, *Mol. Cell Biol.* 16, 1401–1409.

Liou, H. et al. (1994), Sequential induction of NF-kB/Rel family proteins during B-cell terminal differentiation, *Mol. Cell Biol.* 14, 5349–5359.

Little, A., and O'Callaghan, J. (1999), TNF-alpha, IL-1alpha, and IL-1beta gene expression is not altered in response to trimethyltin-induced neuronal damage in the adult rat hippocampus, *Soc. Neurosci. Abs.* 25, 1535.

Little, A. J., and O'Callaghan, J. (2001), The Astrocyte Response to neural injury: a review and reconsideration of key features, *Site-Specific Neurotoxicity*, ed. R. D. Leste et al. In press. Harwood Academic, Amsterdam.

Little, A. et al. (2001), Hippocampal injury in the absence of a compromised blood-brain barrier (BBB) results in enhanced microglial expression of MCP-1 without affecting other proinflammatory mediators, *Soc. Neurosci. Abs.* 439.2.

Maier, W. E. et al. (1997), Induction of tumor necrosis factor alpha in cultured glial cells by trimethyltin, *Neurochem. Int.* 30, 385–392.

Makarov, S. S. (2000), NF-kappaB as a therapeutic target in chronic inflammation: recent advances, *Mol. Med. Today* 6, 441–448.

Mao, X., and Barger, S. W. (1998), Neuroprotection by dehydroepiandrosterone-sulfate: role of an NF-kappaB- like factor, *Neuroreport* 9, 759–763.

Mattson, M., and Camandola, S. (2001), NF-kB in neuronal plasticity and neurodegenerative disorders, *J. Clinical Invest.* 107, 247–254.

Mattson, M. et al. (2000), Roles of nuclear factor kappa B in neuronal survival and plasticity, *J. Neurochem.* 74, 443–456.

Mattson, M. P., and Chan, S. L. (2001), Dysregulation of cellular calcium homeostasis in Alzheimer's disease: bad genes and bad habits, *J. Mol. Neurosci.* 17, 205–224.

Mattson, M. P., Keller, J. N., and Begley, J. G. (1998), Evidence for synaptic apoptosis, *Exp. Neurol.* 153, 35–48.

Mattson, M. P. et al. (2000), Roles of nuclear factor kappaB in neuronal survival and plasticity, *J. Neurochem.* 74, 443–456.

McCann, M. J. et al. (1996), Differential activation of microglia and astrocytes following trimethyl tin-induced neurodegeneration, *Neuroscience* 72, 273–281.

McKay, L. I., and Cidlowski, J. A. (1999), Molecular control of immune/inflammatory responses: interactions between nuclear factor-kappa B and steroid receptor-signaling pathways, *Endocr. Rev.* 20, 435–459.

McMillan, D. E., and Wenger, G. R. (1985), Neurobehavioral toxicology of trialkyltins, *Pharmacol. Rev.* 37, 365–379.

Meberg, P. et al. (1996), Gene expression of the transcription factor NF-kB in hippocampus: regulation by synaptic activity, *Mol. Brain Res.* 38, 179–190.

Mercurio, F. et al. (1997), IKK-1 and IKK-2: cytokine-activated IkB kinases essential for NF-kB activation, *Science* 278, 860–866.

Mercurio, F. et al. (1999), IkappaB kinase (IKK)-associated protein 1, a common component of the heterogeneous IKK complex, *Mol. Cell Biol.* 19, 1526–1538.

Miyamoto, S., and Verma, I. M. (1995), Rel/NF-kappa B/I kappa B story, *Adv. Cancer Res.* 66, 255–292.

Miyamoto, S., Chiao, P., and Verma, I. (1994), Enhanced IkBa degradation is responsible for constitutive NF-kB activity in mature murine B-cell lines, *Mol. Cell Biol.* 14, 3276–3282.

Monnet-Tschudi, F. et al. (1995), Effects of trimethyltin (TMT) on glial and neuronal cells in aggregate cultures: dependence on the developmental stage, *NeuroToxicology* 16, 97–104.

Moore, P., Ruben, S., and Rosen, C. (1993), Conservation of transcriptional activation functions of the NF-kB p50 and p65 subunits in mammalian cells and *Saccharomyces cerevisiae*, *Mol. Cell Biol.* 13, 1666–1674.

Mora, A. et al. (2001), NF-kappa B/Rel participation in the lymphokine-dependent proliferation of T lymphoid cells, *J. Immunol.* 166, 2218–2227.

Morgan, J. I. et al. (1987), Mapping patterns of c-fos expression in the central nervous system after seizure, *Science* 237, 192–197.

Morolanu, J. (1998), Distinct nuclear import and export pathways mediated by members of the karyopherin beta family, *Cell Biochem.* 70, 231–239.

Mou, S. S. et al. (2002), Myocardial inflammatory activation in children with congenital heart disease, *Crit. Care Med.* 30, 827–832.

Muller, C., and Harrison, S. (1995), The structure of the NF-kB p50:DNA complex: a starting point for the Rel family, *FEBS Lett.* 369, 113–117.

Muller, J. M., Ziegler-Heitbrock, H. W., and Baeuerle, P. A. (1993), Nuclear factor kappa B, a mediator of lipopolysaccharide effects, *Immunobiology* 187, 233–256.

Muller, M., and Sorrell, T. C. (1997), Oxidative stress and the mobilisation of arachidonic acid in stimulated human platelets: role of hydroxyl radical, *Prostaglandins* 54, 493–509.

Nabel, G., and Baltimore, D. (1987), An inducible transcription factor activates expression of human immunodeficiency virus in T cells, *Nature* 326, 711–713.

Nakshatri, H. et al. (1997), Constitutive actions of NF-kB during progression of breast cancer to hormone-independent growth, *Mol. Cell Biol.* 17, 3629–3639.

Neve, B. P., Fruchart, J. C., and Staels, B. (2000), Role of the peroxisome proliferator-activated receptors (PPAR) in atherosclerosis, *Biochem. Pharmacol.* 60, 1245–1250.

Nishimura, T. et al. (2001), Changes in the GABA-ergic system induced by trimethyltin application in the rat, *Brain Res. Mol. Brain Res.* 97, 1–6.

Nishimura, T. et al. (2001), Changes in the expression of GABA receptors after trimethyltin in the rat, *Soc. Neurosci. Abs.* 971.15.

Nonaka, M. et al. (1999), Prolonged activation of NF-kappaB following traumatic brain injury in rats, *J. Neurotrauma* 16, 1023–1034.

Okazaki, M., Evenson, D., and Nadler, J. (1995), Hippocampal mossy fiber sprouting and synapse formation after status epilepticus in rats: visualization after retrograde transport of biocytin, *J. Comp. Neurol.* 352, 515–534.

Patel, M. et al. (1990), Interaction of trimethyltin with hippocampal glutamate, *NeuroToxicology* 11, 601–608.

Patterson, S. L. et al. (1992), Neurotrophin expression in rat hippocampal slices: a stimulus paradigm inducing LTP in CA1 evokes increases in BDNF and NT-3 mRNAs, *Neuron* 9, 1081–1088.

Pemberton, L., Blobel, G., and Rosenblum, J. (1998), Transport routes through the nuclear pore complex, *Curr. Opin. Cell Biol.* 10, 392–399.

Penkowa, M. et al. (2001), Zinc or copper deficiency-induced impaired inflammatory response to brain trauma may be caused by the concomitant metallothionein changes, *J. Neurotrauma* 18, 447–463.

Pennypacker, K. et al. (2001), NF-kB p50 is increased in neurons surviving hippocampal injury, *Exp. Neurol.* 172, 307–319.

Perez-Otaño, I. et al. (1996), Induction of NF-kB-like transcription factors in brain areas susceptible to kainate toxicity, *Glia* 16, 306–315.

Philbert, M. A. et al. (2000), Mechanisms of injury in the central nervous system, *Toxicol. Pathol.* 28, 43–53.

Pierce, J., Lenardo, M., and Baltimore, D. (1988), Oligonucleotide that binds nuclear NF-kB acts as a lymphoid-specific and inducible enhancer element, *Proc. Natl. Acad. Sci. USA* 85, 1482–1486.

Rong, Y., and Baudry, M. (1996), Seizure activity results in a rapid induction of nuclear factor-kappa B in adult but not juvenile rat limbic structures, *J. Neurochem.* 67, 662–668.

Ross, W. D. et al. (1981), Neurotoxic effects of occupational exposure to organotins, *Am. J. Psychiatry* 138, 1092–1095.

Rostasy, K. et al. (2000), NF-kappaB activation, TNF-alpha expression, and apoptosis in the AIDS-Dementia-Complex, *J. Neurovirol.* 6, 537–543.

Rothwarf, D., and Karin, M. (1999), The NF-kappa B activation pathway: a paradigm in information transfer from membrane to nucleus, *Sci. STKE* 1999, RE1.

Rothwarf, D. et al. (1998), IKKg is an essential regulatory subunit of the IkappaB kinase complex, *Nature* 395, 297–300.

Sachdev, S., and Hannink, M. (1998), Loss of IkBa-mediated control over nuclear import and DNA binding enables oncogenic activation of c-Rel, *Mol. Cell Biol.* 18, 5445–5456.

Salminen, A., Liu, P., and Hsu, C. (1995), Alteration of transcription factor binding activities in the ischemic rat brain, *Biochem. Biophys. Res. Commun.* 212, 939–944.

Savory, J. et al. (1999), Age-related hippocampal changes in Bcl-2:Bax ratio, oxidative stress, redox-active iron and apoptosis associated with aluminum-induced neurodegeneration: increased susceptibility with aging, *NeuroToxicology* 20, 805–817.

Schmidt-Ullrich, R. et al. (1996), NF-kB activity in transgenic mice: developmental regulation and tissue specificity, *Development* 122, 2117–2128.

Schmitz, M. L. et al. (1994), Structural and functional analysis of the NF-kappa B p65 C terminus. An acidic and modular transactivation domain with the potential to adopt an alpha-helical conformation, *J. Biol. Chem.* 269, 25613–25620.

Schütze, S. et al. (1992), TNF activates NF-kB by phosphatidylcholine-specific phospholipase C-induced "acidic" sphingomyelin breakdown, *Cell* 71, 765–766.

Sen, R., and Baltimore, D. (1986), Inducibility of kappa immunoglobulin enhancer-binding protein NF-kappa B by a posttranslational mechanism, *Cell* 47, 921–928.

Sha, W. et al. (1995), Targeted disruption of the p50 subunit of NF-kB leads to multifocal defects in immune responses, *Cell* 80, 321–330.

Sharp, F. R. et al. (2000), Multiple molecular penumbras after focal cerebral ischemia, *J. Cereb. Blood Flow Metab.* 20, 1011–1032.

Siebenlist, U., Franzoso, G., and Brown, K. (1994), Structure, regulation and function of NF-kB, *Annu. Rev. Cell Biol.* 10, 405–455.

Smith-Swintosky, V. et al. (1994), Secreted forms of beta-amyloid precursor protein protect against ischemic brain injury, *J. Neurochem.* 63, 781–784.

Stine, K. E., Reiter, L. W., and Lemasters, J. J. (1988), Alkyltin inhibition of ATPase activities in tissue homogenates and subcellular fractions from adult and neonatal rats, *Toxicol. Appl. Pharmacol.* 94, 394–406.

Tam, W. et al. (2000), Cytoplasmic sequestration of Rel proteins by IkBa requires CRM1-dependent nuclear export, *Mol. Cell Biol.* 20, 2269–2284.

Tam, W. F., and Sen, R. (2001), IkappaB family members function by different mechanisms, *J. Biol. Chem.* 276, 7701–7704.

Terai, K. et al. (1996), Enhancement of immunoreactivity for NF-kappa B in human cerebral infarctions, *Brain Res.* 739, 343–349.

Terai, K., Matsuo, A., and McGeer, P. (1996), Enhancement of immunoreactivity for NF-kB in the hippocampal formation and cerebral cortex of Alzheimer disease, *Brain Res.* 735, 159–168.

Toggas, S. M., Krady, J. K., and Billingsley, M. L. (1992), Molecular neurotoxicology of trimethyltin: identification of stannin, a novel protein expressed in trimethyltin-sensitive cells, *Mol. Pharmacol.* 42, 44–56.

Toillon, R. A. et al. (2002), Normal breast epithelial cells induce apoptosis of breast cancer cells via Fas signaling, *Exp. Cell Res.* 275, 31–43.

Traenckner, E. B., Wilk, S., and Baeuerle, P. A. (1994), A proteasome inhibitor prevents activation of NF-kappa B and stabilizes a newly phosphorylated form of I kappa B-alpha that is still bound to NF-kappa B, *EMBO J.* 13, 5433–5441.

Urban, M. B., and Baeuerle, P. A. (1991), The role of the p50 and p65 subunits of NF-kappa B in the recognition of cognate sequences, *New Biol.* 3, 279–288.

Urban, M. B., Schreck, R., and Baeuerle, P. A. (1991), NF-kappa B contacts DNA by a heterodimer of the p50 and p65 subunit, *EMBO J.* 10, 1817–1825.

Vaccari, A. et al. (1997), Is increased neurotoxicity a burden of the ageing brain? *Adv. Exp. Med Biol.* 429, 221–234.

Van Kessel, K. P. et al. (1987), Further evidence against a role for toxic oxygen products as lytic agents in NK cell-mediated cytotoxicity, *Immunology* 62, 675–678.

Verma, I. et al. (1995), Rel/NF-kB/IkB family: intimate tales of association and dissociation, *Genes Dev.* 9, 2723–2735.

Viviani, B. et al. (1998), Glia increase degeneration of hippocampal neurons through release of tumor necrosis factor-alpha, *Toxicol. Appl. Pharmacol.* 150, 271–276.

Viviani, B. et al. (2001), Trimethyltin-activated cyclooxygenase stimulates tumor necrosis factor-alpha release from glial cells through reactive oxygen species, *Toxicol. Appl. Pharmacol.* 172, 93–97.

Weber, G. F. (1999), Final common pathways in neurodegenerative diseases: regulatory role of the glutathione cycle, *Neurosci. Biobehav. Rev.* 23, 1079–1086.

Whiteside, S., and Israël, A. (1997), I kappa B proteins: structure, function and regulation, *Semin. Cancer Biol.* 8, 93–101.

Winston, J. et al. (1999), The SCF b-TCP-ubiquitin ligase complex associates specifically with phosphorylated destruction motifs in IkBa and b-catenin and stimulates IkBa ubiquitination *in vitro. Genes Dev.* 13, 270.

Xia, Z. et al. (1995), Opposing effects of ERK and JNK-p38 MAP kinases on apoptosis, *Science* 270, 1326–1331.

Yan, S. et al. (1996), RAGE and amyloid-beta peptide neurotoxicity in Alzheimer's disease, *Nature* 382, 685–691.

Yang, K., Mu, X. S., and Hayes, R. L. (1995), Increased cortical nuclear factor-kappa B (NF-kappa B) DNA binding activity after traumatic brain injury in rats, *Neurosci. Lett.* 197, 101–104.

Yaron, A. et al. (1998), Identification of the receptor component of the IkappaBalpha-ubiquitin ligase, *Nature* 396, 590–594.

Yu, Z. et al. (1999), Lack of p50 subunit of nuclear factor-kB increases the vulnerability of hippocampal neurons to excitotoxic injury, *J. Neurosci.* 19, 8856–8865.

Yu, Z. et al. (2000), Neuroprotective role for the p50 subunit of NF-kappaB in an experimental model of Huntington's disease, *J. Mol. Neurosci.* 15, 31–44.

Zabel, U. et al. (1993), Nuclear uptake control of NF-kB by MAD-3, an IkB protein present in the nucleus, *EMBO J.* 12, 201–211.

Zandi, E. et al. (1997), The I kappaB kinase complex (IKK) contains two kinase subunits, IKKa and IKKb, necessary for IkB phosphorylation and NF-kB activation, *Cell* 91, 243–252.

CHAPTER 5

Functional Implications of NMDA Receptor Subunit Expression in the Rat Brain Following Developmental Pb^{2+} Exposure

Tomás R. Guilarte

CONTENTS

5.1 DEVELOPMENTAL LEAD EXPOSURE: A GLOBAL PROBLEM

Lead (Pb^{2+}) is a ubiquitous environmental neurotoxicant with a long history of exposure in children (Needleman, 1998; Mielke, 1999). The developing brain is highly susceptible to Pb^{2+} exposure and long-term deficits in cognitive function are the principal effects of Pb^{2+}-induced neurotoxicity (Bellinger et al., 1987; McMichael et al., 1988; Bellinger et al., 1992; Winneke et al., 1994; Lanphear et al.,

041528031-1/04/$0.00+$1.50

2000). Despite significant efforts during the last two decades to reduce Pb^{2+} levels in the environment, one in twenty children living in the United States exhibit blood Pb^{2+} levels (>10 μg/dL) known to produce long-term deficits in cognitive function (Pirkle et al., 1994; 1998). The most prominent route of exposure at the present time is from Pb^{2+} found in the dust and soils in homes that were built prior to the 1970s in which leaded paint was used (Lanphear et al., 1999). A recent report by the United States Surgeon General indicates that "lead poisoning poses one of the greatest environmental threats to children in America" (Satcher, 2000). This environmental problem is not exclusive to the United States; in fact, emerging data from other industrialized and developing nations suggest that many more children are exposed to even higher levels of Pb^{2+} from contamination of their living environment (Lopez-Carrillo et al., 1996; Romieu et al., 1997; Factor-Litvak et al., 1999; Kaul et al., 1999; Gao et al., 2001; Rubin et al., 2002). Therefore, the effect of Pb^{2+} exposure on the cognitive development of children is a worldwide environmental health problem with significant social and economic consequences (Grosse et al., 2002; Landrigan et al., 2002). The overwhelming evidence of this environmental catastrophe underscores the need for understanding the molecular bases of Pb^{2+}-induced learning impairments in order to devise useful intervention strategies to ameliorate its neurotoxic effects.

Animal models have been useful in characterizing the learning and memory impairments (Winneke et al., 1977; Rice, 1984; 1990; Cory-Slechta, 1995; Jett et al., 1997; Kuhlmann et al., 1997) associated with developmental Pb^{2+} exposure as well as the cellular and molecular mechanisms underpinning the cognitive dysfunction (Nihei and Guilarte, 2001). This chapter will focus on the knowledge gained during the last decade on the interaction of Pb^{2+} with a unique subtype of glutamatergic receptors, the N-methyl-D-aspartate receptor (NMDAR), as a molecular target responsible for some of the cognitive deficits associated with developmental Pb^{2+} neurotoxicity. Further, I will focus the discussion on NMDAR in the hippocampus, since this is a brain region containing high concentrations of NMDAR, it is important in learning and memory, and it is highly sensitive to the neurotoxic effects of developmental Pb^{2+} exposure.

5.2 NMDA RECEPTOR FUNCTION IN SYNAPTIC PLASTICITY AND COGNITIVE FUNCTION

Glutamate is the major excitatory neurotransmitter in the mammalian brain and mediates activity-dependent processes critical to both the developing and the mature brain. Ionotropic and metabotropic receptor subtypes mediate the actions of glutamate, and activation of the ionotropic NMDAR plays a central role in brain development, learning and memory as well as in neurodegenerative diseases (Collingridge and Lester, 1989; Scheetz and Constantine-Paton, 1994; Ozawa et al., 1998). The NMDAR has unique characteristics that conform to those postulated in the Hebbian model of synaptic plasticity. It has long been hypothesized that learning and consolidation of memories in the mammalian brain involves the modification of synapses. Hebb's postulate promoting the idea of "correlated activity" as a cellular basis of learning and memory suggests that when presynaptic and postsynaptic neurons are active simultaneously, their connections are strengthened (Hebb, 1949). The NMDAR behaves within the confines of this Hebbian model in that it acts as a coincident detector requiring the simultaneous binding of the endogenous agonist glutamate (released from presynaptic neurons) and the depolarization of the neuronal membrane (postsynaptic neuron) in order for the receptor to be activated. If these two events occur simultaneously, the magnesium block of the NMDAR associated ion channel is removed allowing for the influx of calcium (Nowak et al., 1984). Therefore, it appears that the NMDAR can implement Hebb's postulate at the synaptic level.

Activity-dependent change in intracellular calcium levels mediated by activation of NMDAR induces a cascade of signal transduction events essential for modification of synapses. Among synaptic processes that are NMDAR dependent, long-term potentiation (LTP) provides an important model of learning and memory. Long-term potentiation was first described and extensively studied in the hippocampus as a long-lasting increase in synaptic efficacy following brief periods of stimulation of specific synapses (Bliss and Lomo, 1973). In the hippocampus, glutamatergic Schaffer–collateral–CA1 and perforant path–dentate gyrus synapses have been the most widely studied (Teyler and DiScenna, 1987; Madison et al., 1991; Malenka and Nicoll, 1999), although, LTP has also been described in a number of other brain regions (Tsumoto, 1992). The hippocampus is a brain region important in the acquisition and consolidation of some forms of learning and memory, especially those involved in place learning or spatial learning. The disruption of hippocampal function by a variety of methods produces deficits in behavioral paradigms that assess these forms of learning and memory (Izquierdo, 1993; McNamara and Skelton, 1993).

It is now widely accepted that some form of LTP in the hippocampus, specifically those induced in Schaffer–collateral–CA1 and perforant path–dentate gyrus synapses, are dependent upon NMDAR activation (Zalusky and Nicoll, 1990; Malenka and Nicoll, 1993). Disruption of NMDAR function by pharmacological means (Morris et al., 1986; Parada-Turska and Turski, 1990; Walker and Gold, 1991; Kentros et al., 1998) or by deletion of specific NMDAR subunits using gene knockout techniques (Sakimura et al., 1995; Tsien et al., 1996; Shimizu et al., 2000; Nakazawa et al., 2002) are associated with disruption of hippocampal LTP and learning and memory. Therefore, during the last decade, a vast number of studies have provided compelling evidence linking the NMDAR as playing an essential role in synaptic plasticity and learning and memory.

5.3 DEVELOPMENTAL EXPRESSION OF NMDA RECEPTOR SUBUNIT GENES

Molecular cloning initially identified five genes encoding for two families of homologous NMDAR subunits, the NR1 and NR2A-D (Moriyoshi et al., 1991; Monyer et al., 1992). More recently, another NMDAR subunit, NR3 (previously called NR-L or NR-x-1), has also been cloned and characterized (Ciabarra et al.,

1995; Sucher et al., 1995; Das et al., 1998). The NR1 subunit gene is alternatively spliced at three exons resulting in eight splice variants (Laurie and Seeberg, 1994; Zukin and Bennett, 1995). Alternatively spliced exon 5 at the N-terminus (N-cassette) encodes for a 21 amino acid sequence. Splicing of exon 5 results in transcripts designated as lacking (NR1-a) or containing (NR1-b) the N-cassette [19]. Exons 21 and 22 encode for two C-terminus cassettes, C1 and C2, that code for 37 and 38 amino acid sequences, respectively. The individual splicing of the C1 or C2 cassette results in transcripts designated as NR1-2 and NR1-3. The presence or absence of both C-terminus cassettes results in NR1-1 and NR1-4 variants. The deletion of the C2 cassette alters the reading frame resulting in the creation of additional coding regions of 22 amino acids, the C2' cassette.

The NR1 subunit is obligatory for functional NMDAR *in vivo* and it aggregates with distinct NR2 subunits to form functional NMDAR complexes (Hollmann and Heinemann, 1994; Laube et al., 1998). Of the NR2 subunits, the NR2A and NR2B are most abundantly expressed in hippocampal and cortical tissue (Monyer et al., 1994). The type and number of NR2 subunit that co-aggregates with the NR1 influences the physiological and pharmacological properties of the receptor (Monyer et al., 1992; 1994; Honer et al., 1998; Vicini et al., 1998). The NR3 subunit is also alternatively spliced and can be present in two isoforms (Sun et al., 1998). The NR3 appears to have a regulatory role on NMDAR activity and influences spine density formation in early development (Das et al., 1998). Distinct NMDAR subunits have unique anatomical and developmental expression providing heterogeneity of NMDAR subtypes.

Developmental changes in the mRNA expression of NMDAR subunits in the rodent hippocampus have been documented using *in situ* hybridization or RNase protection assays. For the most part, NR1, NR2A, and NR2B subunits are the most abundantly expressed in the hippocampus (Monyer et al., 1994; Riva et al., 1994; Luo et al., 1996). NR3 subunit expression is also present in the hippocampus during embryonic development and in the early postnatal period, but its expression markedly decreases by 2 weeks of age (Ciabarra et al., 1995; Sucher et al., 1995). Expression of the mRNA and protein for the obligatory NR1 and for the NR2B subunits are present at birth and increase as a function of age to reach maximal levels at 21 to 28 days after birth (Riva et al., 1994; Luo et al., 1996). In the rodent brain at birth, there are no detectable levels of NR2A subunit mRNA, suggesting that during the first week of life the majority of NMDAR complexes in the rat hippocampus have an NR1/NR2B composition (Williams et al., 1993; Riva et al., 1994; Monyer et al., 1994; Zhong et al., 1995). However, since the NR3 subunit is present at this time, it is likely that NR1/NR2B/NR3 receptors may also be expressed. Functional NMDAR having this pairing of subunit composition have been described (Das et al., 1998) and it appears that the insertion of the NR3 subunit decreases the magnitude of NMDAR currents relative to NR1/NR2B containing receptors (Das et al., 1998).

Following the first week of life, there is a rapid increase in the expression of the NR2A subunit gene in the rodent brain. On the other hand the NR2B subunit mRNA remains unchanged or slightly decreased during the first 4 weeks of postnatal development. The emergence of the NR2A subunit is coincident with changes in the physiological properties of NMDAR in the hippocampus (Williams et al., 1993) and there is the appearance of distinct NMDAR subtypes based on biophysical and pharmacological characteristics (Kew et al., 1998; Stocca et al., 1998). The subunit composition of NMDAR complexes during development is further complicated by the fact that the gene of the obligatory NR1 subunit is alternatively spliced and each variant has differential expression during development. Therefore, the potential number of NR1 splice variant combinations with NR2 and NR3 subunits provides heterogeneity of function of NMDAR complexes during development and in the adult brain.

5.4 ACTIVITY-DEPENDENT REGULATION OF NMDA RECEPTOR SUBUNIT EXPRESSION

There is compelling experimental evidence that the expression of NMDAR subunits is under the control of genetically programmed development of excitatory circuitry and as a result of experience-dependent modification of synaptic activity. Neuronal activity during development transforms unstable synapses into topographically organized stable neuronal circuits. A model of activity-dependent NMDAR synaptic plasticity has been described in the visual cortex. The onset and end of the critical period of ocular dominance in the visual cortex can be modified by the amount of light exposure the animal receives (Carmignoto and Vicini, 1992; Quinlan et al., 1999a,b) and it can alter the normal developmental pattern of NMDAR subunit expression (Quinlan et al., 1999a,b). Animals reared in the dark or pharmacological manipulation of neuronal activity in the visual cortex prevents the normal developmental changes in NMDAR channel kinetics (Carmignoto and Vicini, 1992; Quinlan et al., 1999a,b) as well as the pattern of NMDAR subunit expression (Carmignoto and Vicini, 1992; Quinlan et al., 1999a,b). Specifically, it appears that in the absence of visual experience (dark rearing) the ontogenetically defined developmental increase in NR2A subunit expression is diminished and can be restored following exposure to light (Quinlan et al., 1999a,b).

In the rat hippocampus, the period of peak LTP induction occurs at approximately 14–15 days of age (Harris and Teyler, 1984). This period coincides with the time in which hippocampal NMDAR levels are expressed at the highest levels (Hattori et al., 1990). Further, this is a developmental time in which the insertion of NR2A subunit expression is increasing and NR3 subunit contribution decreases with no appreciable difference in NR2B expression. Recent progress in the use of genetically engineered mice with anatomically selective deletion of specific NMDAR subunits has provided valuable information on their role in hippocampal physiology and in learning and memory.

Consistent with *in vivo* studies, the use of hippocampal and cortical neuron cultures provided evidence of activity-dependent modifications in NMDAR subunit expression. Hoffman et al. (2000) have shown that the developmental increase in NR2A subunit is dependent upon calcium influx mediated by NMDAR and L-type calcium channels since pharmacological blockade of NMDAR and L-type calcium channels decreased NR2A subunit expression. On the other hand, the expression of the NR2B subunit was not altered under the same conditions. The decrease in the normal developmental expression of the NR2A subunit produced by antagonism of NMDAR and L-type calcium channels resulted in functional changes in the NMDAR expressed. These investigators showed that following blockade, NMDAR expressed slower deactivation kinetics of NMDA-mediated responses and increased sensitivity to ifenprodil relative to control cultures (Hoffmann et al., 2000). These characteristics are similar to recombinant NMDAR complexes of the NR1/NR2B subtype (Williams, 1993; Williams et al., 1993). Thus, the data suggest that the reduced expression of NR2A subunit by blockade of synaptic activity increases the proportion of NR2B-containing NMDAR complexes. Other studies support this conclusion by showing that a reduction in presynaptic exocytosis regulates NMDAR subunit expression in a similar manner as pharmacological blockade of NMDAR and L-type calcium channels (Hoffmann et al., 1997; Lindlbauer et al., 1998).

5.5 SYNAPTIC TARGETING OF NMDA RECEPTOR SUBUNITS

Synaptic neurotransmission requires the precise localization of receptors in the postsynaptic membrane in direct apposition to neurotransmitter release sites. Excitatory synapses in the mammalian brain are primarily localized in dendritic spines that express postsynaptic densities (PSD). The PSD is a specialized structure within the postsynaptic membrane where neurotransmitter receptors, signaling enzymes and cytoskeletal proteins are clustered (Ziff, 1997; Kennedy, 1998). Activity-dependent delivery of newly synthesized NMDAR subunits to the synapse requires a number of steps including transcription, translation, and post-translational modification of the subunit protein as well as the appropriate targeting and clustering to the PSD. Recent progress has led to a better understanding of mechanisms involved in NMDAR subunit subcellular localization, trafficking, and synaptic targeting. For example, the NR2A and NR2B subunit of the NMDAR and some of the NR1 splice variants are known to directly interact with the PSD-95/SAP-90 family of proteins, a major component of the PSD, in order to anchor and stabilize NMDAR complexes at synaptic sites (Kornau et al., 1995; O'Brien et al., 1998).

Ehlers et al. (1998) showed that NR1 splice variants containing the first carboxy terminus cassette (i.e., the C1 cassette) localized to discrete, receptor-rich regions in the plasma membrane, while splice variants lacking the C1 cassette express homogenous distribution throughout the cytoplasm when transfected into fibroblasts. The C1 cassette also contains consensus sequences for phosphorylation by PKC. In the same study, these investigators show that phosphorylation of serine residues within the C1 cassette resulted in a disruption of the receptor-rich domains in the plasma membrane. An interaction of the C1 cassette with neuronal intermediate filaments was shown in a later study by the same group of investigators (Ehlers et al., 1998) providing a linkage of its interaction with anchoring and cytoskeletal proteins. Other investigations have also provided associations of NMDAR subunits with spectrin (Wechsler and Teichberg, 1998), alpha-actinin-2 (Wyszynski et al., 1998), and Yotiao (Lin et al., 1998), proteins that link neuronal proteins to the cytoskeleton. Interestingly, the interaction of Yotiao with the NR1 subunit is dependent upon the presence of the C1 cassette (Lin et al., 1998), and the interaction of the NR1 subunit with spectrin is inhibited by PKC phosphorylation (Wechsler and Teichberg, 1998).

A more recent study has examined the effect of NR1 splicing on cell surface expression and the formation of functional NMDAR complexes when NR1 isoforms are cotransfected with NR2 subunits (Okabe et al., 1999). Their findings indicate that in several cell types including hippocampal neurons, the NR1 splice isoforms with the longest C-terminal cytoplasmic tail (i.e., NR1-1) express no detectable levels of the protein in the cell surface. On the other hand, NR1 isoforms with the shortest C-terminal tail (i.e., NR1-4) have the greatest cell surface expression (Okabe et al., 1999). Interestingly, the cotransfection of the NR1-4 isoform with NR1-1 increased the NR1-1 cell surface expression suggesting that different splice isoforms are able to interact with each other in intracellular organelles such as the endoplasmic reticulum (ER) to modify their targeting to the cell surface. In the same study it was shown that the co-expression of the NR2B subunit can increase the proportion of cells demonstrating surface expression of the NR1-1 and NR1-2 but not of the NR1-3 and NR1-4 (Okabe et al., 1999). The differential expression of NR1 isoforms in the cell surface was correlated with increases in intracellular calcium concentrations when cells were co-expressed with the NR2B subunit and stimulated with glutamate. Therefore, NR1-4/NR2B receptors had the highest increases in intracellular calcium concentrations and NR1-1/NR2B receptor the lowest (Okabe et al., 1999). Based on these studies the investigators conclude that the differential expression of NR1 splice variants in the cell surface is a critical factor regulating the amount of NMDAR mediated calcium influx by glutamate stimulation.

Another important step in the trafficking of NMDAR to synapses is the exit of newly assembled receptors from the endoplasmic reticulum (ER). A recent study has shown that NR1 splice variants containing the C1 cassette have a ER retention/retrival motif that maintains the NR1 subunit in the ER while splice variants lacking the C1 cassette are efficiently expressed in the plasma membrane (Scott et al., 2001). Further, PKC phosphorylation of the C1 and the introduction of an alternatively spliced consensus PDZ binding domain present in the C2' suppresses ER retention and promotes the trafficking to the plasma membrane. Other studies have shown that transfection of NR1 variants containing the C1 cassette are retained in the ER but co-expression with the NR2A subunit reduces ER localization with enhanced cell surface expression (McIlhinney et al., 1996; 1998). In general, these studies document the importance of NR1 mRNA alternative splicing, PKC phosphorylation and association with NR2 subunits on the trafficking, subcellular localization, and anchoring of functional NMDAR complexes at synaptic sites.

5.6 EFFECT OF DEVELOPMENTAL EXPOSURE TO LEAD ON NMDA RECEPTOR SUBUNIT EXPRESSION

The previous sections have provided an introductory overview of the molecular biology and functional aspects of NMDAR in synaptic plasticity and in learning and memory in the mammalian brain. This basic knowledge provides important insights in understanding the functional consequences associated with developmental Pb^{2+}

exposure and effects on neuronal processes that are NMDAR dependent. In the following section, I will describe what is known about the effects of Pb^{2+} on the developing brain as it relates to the NMDAR and functional correlates that have been characterized in Pb^{2+}-exposed animals.

First, it is important to indicate that Pb^{2+} is a selective and potent NMDAR antagonist with no direct effects on the AMPA/kainate glutamate receptor subtypes (Alkondon et al., 1990; Guilarte and Miceli, 1992; Guilarte et al., 1995). The effect of *in vivo* exposure to Pb^{2+} during development on NMDAR subunit expression has been documented in rodent models of Pb^{2+} neurotoxicity. These studies show that exposure to environmentally relevant levels of Pb^{2+} were effective in producing marked changes in NMDAR subunit gene expression in the developing as well as in the mature brain. The effects of developmental Pb^{2+} exposure on gene expression of NMDAR subunits was first described by Guilarte and McGlothan (1998) using *in situ* hybridization. This work showed that developmental Pb^{2+} exposure produced significant increases in NR1 subunit mRNA expression in selective regions of the rat hippocampus primarily at 14 and 21 days of age. On the other hand, Pb^{2+} exposure produced marked reductions in NR2A subunit mRNA in essentially all hippocampal regions examined at the same age range (Guilarte and McGlothan, 1998). The changes in NR1 and NR2A subunit expression were present in the absence of changes in NR2B subunit mRNA. An important observation in this study was that the developmental expression of the NR1, NR2A, NR2B, or NR2C subunits mRNA was not significantly altered in the cerebellum of the same Pb^{2+}-exposed rats indicative of a regional selectivity of the Pb^{2+} effects. The only exception was a small increase in NR1 subunit mRNA in the cerebellum of Pb^{2+}-exposed rats at 28 days of age (Guilarte and McGlothan, 1998). In this report, the oligoprobe used for the NR1 subunit *in situ* hybridization was directed at one of the transmembrane domains of the NR1 subunit gene that is common to all NR1 splice variants. Therefore, in a subsequent study, Guilarte et al. (2000) examined the effects of developmental Pb^{2+} exposure on the NR1 splice variants most abundantly expressed in the rat hippocampus. The findings showed that there were differential effects of Pb^{2+} exposure on NR1 splice variants in the different regions of the hippocampus. In regards to the N-terminal cassette, expression of the NR1a splice variant mRNA was significantly increased throughout the pyramidal and granule cell layers of Pb^{2+}-exposed rats at postnatal days 14 and 21 (Guilarte et al., 2000). Increased levels of NR1b splice variant mRNA was also measured in the pyramidal cell layer at 21 days of age. Therefore, both the NR1a and NR1b transcripts were increased in the hippocampus of Pb^{2+}-exposed rats during development. Developmental Pb^{2+} exposure also altered the splicing of the C-terminus cassettes. The expression of the NR1-2 isoform lacking the C1 cassette was increased in CA4 pyramidal cells and in granule cells of the dentate gyrus at 21 days of age. Further, NR1-4 mRNA also lacking the C1 cassette was increased in the CA3 pyramidal cells at 14 days of age but decreased in the same region at 21 days of age (Guilarte et al., 2000). No effect of Pb^{2+} exposure was measured on NR1-1 splice variant mRNA expression in the same animals.

A recent study on the effects of developmental Pb^{2+} exposure on NMDAR subunit expression has essentially confirmed the findings by Guilarte and colleagues and have extended the work to assess other NMDAR subunits not previously examined

(Zhang et al., 2002). These investigators also found significant changes in NR2D (increased) and NR3 (decreased) mRNA expression in the hippocampus of developing rats exposed to Pb^{2+}. Therefore, Pb^{2+} exposure alters the normal ontogenetic expression of a number of the NMDAR subunits with regional and temporal specificity. The functional consequences of these changes are yet to be unraveled but the changes provide important insights on the effects of Pb^{2+} exposure on the pharmacology and cell surface expression of NMDAR complexes in the rat brain.

5.7 FUNCTIONAL IMPLICATIONS AND CORRELATES OF NMDA RECEPTOR SUBUNIT CHANGES IN DEVELOPMENTAL Pb^{2+} NEUROTOXICITY

Historically, our understanding of effects from developmental exposure to Pb^{2+} at the molecular, cellular, and whole animal level has often occurred in bits and pieces with a concentration on a specific aspect of the puzzle. Our aim during the last decade has been to use a "top-down" approach in which environmentally relevant exposure paradigms in developing animals produces deficits of learning and memory that can then be used to make associations with cellular and molecular changes in the same or in similarly treated animals. Our initial approach was to determine whether developmental exposure to low levels of Pb^{2+} resulted in impairments in a specific behavioral test that assesses spatial learning. We selected the water maze as our initial behavioral paradigm to test the effects of developmental Pb^{2+} exposure because performance in the water maze is known to be dependent on the integrity of the hippocampus and in functional NMDAR complexes. Further, developmental exposure to Pb^{2+} was already known to target the hippocampus and Pb^{2+} was a known inhibitor of the NMDAR complex (Hashemzadeh-Gargari and Guilarte, 1999).

Our initial studies showed that rats that were exposed to Pb^{2+} during development expressed a deficit in performance in the water maze spatial learning task relative to control animals shortly after weaning (postnatal day 21) but not as young adults (56 days of age) (Jett et al., 1997). This initial study indicated that developing rats, similar to children, were more susceptible to Pb^{2+}-induced learning deficits than adult rats. However, we were puzzled that young adult rats that were continuously exposed to Pb^{2+} during development did not exhibit a deficit in performance of the water maze suggestive that maybe the learning deficit was no longer present as the animals aged. We reasoned that one possible explanation may be that the behavioral task used was not sufficiently difficult to detect a difference in performance between adult Pb^{2+}-exposed and control rats. Based on this hypothesis, another series of experiments was performed in which the water maze paradigm was modified to make it a more difficult task. The change was to simply reduce the number of swim trials per day from four trials to one trial. The outcome of this change in water maze performance was indeed a demonstration that young adult rats that were exposed to Pb^{2+} during development in fact were impaired in learning the water maze task relative to controls (Kuhlmann et al., 1997). Therefore, a negative outcome does not prove the lack of an effect and one must apply the appropriate degree of difficulty in a behavioral test in order to be able to demonstrate differences in performance.

Another important point demonstrated by the study of Kuhlmann et al. (1997) was that there was a "developmental window" in which exposure to low levels of Pb^{2+} are able to affect cognitive function and the effects are long-lasting. The study showed that Pb^{2+} exposure during gestation and lactation until weaning resulted in impaired performance in the water maze of adult rats (100 days of age) despite the fact that one group of the Pb^{2+}-exposed animals was removed from the exposure at weaning (21 days of age). Further, if the Pb^{2+} exposure was initiated at weaning and the animals were tested as adult rats (100 days of age), no effect on water maze performance was measured (Kuhlmann et al., 1997). In summary, these studies showed that there is a period during brain development in which neuronal processes are highly vulnerable to the presence of Pb^{2+} in altering the ability of the animal to learn tasks that are dependent upon the normal function of the hippocampus.

Prior to and during the time that these studies were being performed, reports appeared in the literature demonstrating the effects of developmental Pb^{2+} exposure on hippocampal long-term potentiation (LTP). LTP is a form of synaptic plasticity that is believed to form the basis for the encoding of information in the mammalian brain (Malenka and Nicoll, 1999). A number of groups, principally those from Dr. Herbert Wiegand in Germany and Dr. Mary Gilbert in the United States, provide evidence that rats exposed to Pb^{2+} during development exhibited deficits in hippo-campal LTP (Altmann et al., 1993; Lasley et al., 1993; Altmann et al., 1994; Gilbert et al., 1996; Gilbert and Mack, 1998; Gutowski et al., 1998). These findings have been corroborated by a number of other laboratories including our own (Zaiser and Miletic, 1997; Ruan et al., 1998; Nihei et al., 2000).

The studies of hippocampal LTP in Pb^{2+}-exposed rats provided a key piece of the puzzle because the type of hippocampal LTP studied was dependent upon the activation of NMDAR. These findings were consistent with the emerging evidence from molecular studies that the NMDAR was a target for Pb^{2+}-induced effects in the hippocampus. An important but often overlooked observation on the effects of Pb^{2+} on hippocampal LTP was that described by Gutowski in Dr. Wiegand's labo-ratory (Gutowski et al., 1998). They showed that while exposure to low levels of Pb^{2+} during development was able to produce deficits in NMDAR-dependent hippo-campal LTP, the same Pb^{2+} exposure paradigm did not alter hippocampal LTP that was not NMDAR dependent, principally LTP occurring in mossy fiber-CA3 synapses (Gutowski et al., 1998). These seminal studies gave support to the hypothesis that NMDAR was a target for Pb^{2+}-induced alterations in synaptic plasticity and in learning and memory. Further, it provides evidence that the locus of Pb^{2+}-induced effects on hippocampal LTP is postsynaptic and not presynaptic, since both NMDAR-dependent and -independent forms of LTP in the hippocampus require glutamate release. It has been suggested that the effects of Pb^{2+}-exposure on hippo-campal LTP are associated with a reduction in glutamate release documented using *in vivo* microdialysis studies in the rodent brain (Lasley and Gilbert, 1996; Lasley et al., 2001).

To further strengthen an association between Pb^{2+}-induced impairments of learn-ing and LTP with changes in NMDAR subunit mRNA and protein expression in the hippocampus, we performed a study in which we examined all of these parameters in littermates (Nihei et al., 2000). The results showed that impairments of spatial learning and deficits in hippocampal LTP in young adult rats (e.g., 50 days of age) exposed to Pb^{2+} during development was associated with deficits in NR1 and NR2A subunit mRNA and protein expression in the hippocampus. For the most part, there were no changes in NR2B subunit mRNA expression in the hippocampus with the exception of a small but significant increase in NR2B mRNA in the CA3 pyramidal cell layer. These findings make a direct association between molecular changes in NR1 and NR2A subunits of the NMDAR and deficits in hippocampal LTP and impairments in spatial learning (Nihei et al., 2000; Nihei and Guilarte, 2001).

The deficits in NR1 and NR2A subunit expression in the hippocampus of young adult rats exposed to Pb^{2+} during development provide important clues in our understanding of NMDAR-dependent mechanisms that help explain the deficits in hippocampal LTP and spatial learning. The functional changes observed in Pb^{2+}-exposed animals can be compared with those in which selective deletion of specific NMDAR subunits have been performed by genetic manipulation. For example, mice lacking the ε1 subunit (GluRε1), the mouse equivalent of the NR2A in the rat, exhibit a deficit in hippocampal LTP (Sakimura et al., 1995; Kiyama et al., 1998). The impairment in LTP in ε1 mutant mice was attributed to an increased threshold for LTP induction (Kiyama et al., 1998). Further, these mice also express impairment in spatial learning and contextual fear learning (Sakimura et al., 1995; Kiyama et al., 1998). Similarly, developmental Pb^{2+} exposure produces reduced expression of NR2A subunit (Guilarte and McGlothan, 1998; Nihei and Guilarte, 1999; Nihei et al., 2000), an increased threshold for LTP induction (Gilbert et al., 1996) and deficits in spatial learning (Kuhlmann et al., 1997; Nihei et al., 2000). In both the ε1 mutant mice and in Pb^{2+}-exposed rats, the impairment in hippocampal LTP can be overcome if the stimulus intensity used to induce LTP is increased (see Gilbert et al., 1996 and Kiyama et al., 1998). Importantly, another effect of ε1 disruption in mice was a faster decay of the maintenance phase of LTP (Kiyama et al., 1998), an effect that is also present in Pb^{2+}-exposed rats (Gilbert and Mack, 1998; Nihei et al., 2000). Thus, reductions in NR2A subunit expression produced by developmental Pb^{2+} exposure result in a functional phenotype similar to the ε1 mutant mice. Taken together, these findings strongly suggest that Pb^{2+}-induced changes in the subunit composition of NMDAR complexes may be responsible for altering the induction threshold of a physiologically relevant phenomenon (i.e., LTP) thought to be a cellular model for learning and memory. Further, these changes may form the basis for the impairment in spatial learning observed in these animals.

The reduced expression of the NR1 subunit in the hippocampus of Pb^{2+}-exposed rats is also consistent with what is known on the importance of the NR1 subunit in hippocampal LTP and learning and memory. There is a growing body of evidence demonstrating the critical involvement of the NR1 subunit of the NMDAR in hippocampal LTP and in the acquisition and consolidation of learning and in associative memory recall (Thomas et al., 1994; Tsien et al., 1996; Eckles-Smith et al., 2000; Cammarota et al., 2000; Shimizu et al., 2000; Nakazawa et al., 2002). The most recent evidence shows that the regionally selective deletion of the NR1 subunit, for example in CA1 pyramidal cells, results in deficits of hippocampal LTP and impairment in learning and memory (Shimizu et al., 2000, Nakazawa et al., 2002). Therefore, the changes in the expression of NMDAR subunits in the hippocampus of rats exposed to low levels of Pb^{2+} during development produces alterations in synaptic plasticity and in cognitive function that are consistent with those observed when the same subunits are selectively altered by genetic manipulation. This overwhelming evidence points to the NMDAR as a principal target for Pb^{2+}-induced deficits in hippocampal LTP and in learning and memory (Nihei and Guilarte, 2001).

5.8 ARE LEAD-INDUCED DEFICITS OF LEARNING AND MEMORY REVERSIBLE?

There is convincing evidence from the human and experimental animal literature that deficits in cognitive function resulting from exposure to Pb^{2+} during early childhood lasts well into adulthood (White et al., 1993). In fact, the current thinking is that the neurological damage produced by exposure to Pb^{2+} during early life is irreversible. We have recently questioned this dogma by assessing an intervention strategy, "environmental enrichment," that may be helpful in reversing the cognitive and molecular deficits induced by Pb^{2+}. There is a growing body of literature on the effects of environmental enrichment on brain function that is beyond the scope of this review; see van Praag et al. (2000) for a comprehensive review. Briefly, environmental enrichment in the experimental animal literature is defined as a combination of "complex inanimate objects and social stimulation" (van Praag et al., 2000). Enrichment has been demonstrated to enhance LTP in the hippocampus and improve spatial learning performance (Nilsson et al., 1999; van Praag et al., 2000; Duffy et al., 2001). The molecular basis of these changes is not fully known. However, environmental enrichment has been shown to increase the expression of nerve growth factors, neurotransmitter receptors, and granule cell neurogenesis in the hippocampus (van Praag et al., 2000). Based on this knowledge, we asked whether environmental enrichment is a paradigm that could reverse the cognitive changes induced by exposure to low levels of Pb^{2+} during development. In this study, we used an experimental paradigm in which the animals were exposed to Pb^{2+} during gestation and lactation until weaning at which time they were removed from the Pb^{2+} exposure. We had previously shown that this paradigm resulted in spatial learning deficits when animals were tested for learning performance as adults (Kuhlmann et al., 1997; Nihei et al., 2000). Our findings indicate that Pb^{2+}-exposed animals placed in "enrichment cages" immediately after the Pb^{2+}-exposure were able to learn better than Pb^{2+}-exposed animals placed in standard laboratory cages and equally as well as control animals placed in the enriched environment. Therefore, environmental enrichment was able to reverse the learning deficits induced by Pb^{2+} (Guilarte et al., 2003). Further, the recovery in learning performance was associated with a recovery of deficits in NR1 subunit gene expression in the hippocampus of Pb^{2+}-exposed rats. The reversal in the cognitive and NR1 subunit gene expression deficits by environmental enrichment was present in the absence of changes in NR2B, PSD-95, GluR1, and αCamKII gene expression pointing to the importance of the NR1 subunit gene in the learning impairments induced by Pb^{2+} (Guilarte et al., 2003). This study is the first demonstration that the cognitive deficits induced by exposure to low levels of Pb^{2+} during development are reversible. Further, they provide for the first time an intervention strategy that may be beneficial to improve the lives of the millions of children who suffer from the consequences of Pb^{2+} exposure.

5.9 FUTURE STUDIES

The studies presented in this chapter are a small part of the larger literature on Pb^{2+} effects on the central nervous system. Because of space limitations, we could not accommodate discussion of all published studies that may be relevant to the topic. However, the evidence presented makes it clear that the direct inhibitory action of Pb^{2+} on the NMDAR alters the expression of its own subunits and thus changes the composition of NMDAR complexes expressed in the developing and adult rat brain. The effects of Pb^{2+} on the NMDAR appear to have a temporal and regional specificity that should impart selective effects on intracellular signaling pathways whose activation are dependent on the entry of calcium ions modulated by the opening of the NMDAR channel. We view the effects of Pb^{2+} on the NMDAR as the tip of the iceberg. There is much more work to be done in determining the effects of Pb^{2+} on the vast number of intracellular signaling pathways that are responsible in propagating signals from the extracellular environment via the NMDAR. These signals are obligatory in turning "on" or "off" the expression of genes that are essential for the cell to respond to the ever-changing neuronal environment. In this context, the ability of environmental enrichment to reverse the learning deficits induced by developmental exposure to Pb^{2+} provides a valuable model to characterize the important molecular players responsible for the devastating effects of Pb^{2+} on the developing brain.

ACKNOWLEDGMENTS

I wish to express my appreciation to the outstanding doctoral students and post-doctoral fellows who have been an integral part of this research. In particular, I would like to thank Dr. David Jett, Dr. Michelle Nihei, Dr. Anthony Kuhlmann, and Christopher Toscano. I also wish to thank Reneé Micelli and Jennifer McGlothan for the meticulous work that they have performed as part of my laboratory throughout the years. This work would not have been possible without the financial support from a grant from the National Institute of Environmental Health Sciences (ES06189).

REFERENCES

Alkondon, M. et al. (1990), Selective blockade of NMDA-activated channel currents may be implicated in learning deficits caused by lead, *FEBS Lett.* 261, 124–130.
Altmann, L. et al. (1993), Impairment of long-term potentiation and learning following chronic lead exposure, *Toxicol. Lett.* 66, 105–112.

Altmann, L., Gutowski, M., and Wiegand, H. (1994), Effects of maternal lead exposure on functional plasticity in the visual cortex and hippocampus of immature rats, *Brain Res. Dev. Brain Res.* 81, 50–56.

Bellinger, D. et al. (1987), Longitudinal analyses of prenatal and postnatal lead exposure and early cognitive development, *New Engl. J. Med.* 316, 1037–1043.

Bellinger, D.C., Stiles, K.M., and Needleman, H.L. (1992), Low-level lead exposure, intelligence and academic achievement: a long-term follow-up study, *Pediatrics* 90, 855–861.

Bliss, T. V., and Lomo, T. (1973), Long-lasting potentiation of synaptic transmission in the dentate area of the anaesthetized rabbit following stimulation of the perforant path, *J. Physiol.* (London) 232, 331–356.

Cammarota, M. et al. (2000), Rapid and transient learning-associated increase in NMDA NR1 subunit in the rat hippocampus, *Neurochem. Res.* 25, 567–572.

Carmignoto, G., and Vicini, S. (1992), Activity-dependent decrease in NMDA receptor responses during development of the visual cortex, *Science* 258, 1007–1011.

Ciabarra, A.M. et al. (1995), Cloning and characterization of x-1: a developmentally regulated member of a novel class of ionotropic glutamate receptor family, *J. Neurosci.* 15, 6498–6508.

Collingridge, G. L., and Lester, R.A.J. (1989), Excitatory amino acid receptors in the vertebrate central nervous system, *Pharmacol. Rev.* 40, 143–210.

Cory-Slechta, D. A. (1995), Relationships between lead-induced learning impairments and changes in dopaminergic, cholinergic, and glutamatergic neurotransmitter system functions, *Annu. Rev. Pharmacol. Toxicol.* 35, 391–415.

Das, S. et al. (1998), Increased NMDA current and spine density in mice lacking the NMDA receptor subunit NR3A, *Nature* 393, 377–381.

Duffy, S.N. et al. (2001), Environmental enrichment modifies the PKA-dependence of hippocampal LTP and improves hippocampus-dependent memory, *Learn. Mem.* 8, 26–34.

Eckles-Smith, K. et al. (2000), Caloric restriction prevents age-related deficits in LTP and in NMDA receptor expression, *Mol. Brain Res.* 78, 154–162.

Ehlers, M.D., Tingley, W.G., and Huganir, R.L. (1995), Regulated subcellular distribution of the NR1 subunit of the NMDA receptor, *Science* 269, 1734–1737.

Ehlers, M.D. et al. (1998), Splice variant-specific interaction of the NMDA receptor subunit NR1 with neuronal intermediate filaments, *J. Neurosci.* 18, 720–730.

Factor-Litvak, P. et al. (1999), The Yugoslavia prospective study of environmental lead exposure, *Environ. Health Persp.* 107, 9–15.

Gao, W. et al. (2001), Blood lead levels among children aged 1 to 5 years in Wuxi City, China, *Environ. Res.* 87, 11–19.

Gilbert, M.E., Mack, C.M., and Lasley, S.M. (1996), Chronic developmental lead exposure increases the threshold for long-term potentiation in rat dentate gyrus in vivo, *Brain Res.* 736, 118–124.

Gilbert, M.E., and Mack, C.M. (1998), Chronic lead exposure accelerates decay of long-term potentiation in rat dentate gyrus in vivo, *Brain Res.* 789, 139–149.

Grosse, S.D. et al. (2002), Economic gains resulting from the reduction in children's exposure to lead in the United States, *Environ. Health Perspect.* 110, 563–569.

Guilarte, T.R., and Miceli, R.C. (1992), Age-dependent effects of lead on [³H]-MK-801 binding to the NMDA receptor-gated ionophore: *in vitro* and *in vivo* studies, *Neurosci. Lett.* 148, 27–30.

Guilarte, T.R., Miceli, R.C., and Jett, D.A. (1995), Biochemical evidence of an interaction of lead at the zinc allosteric sites of the NMD receptor complex: effects of neuronal development, *NeuroToxicology* 16, 63–72.

Guilarte, T.R., and McGlothan, J.L. (1998), Hippocampal NMDA receptor mRNA undergoes subunit specific changes during developmental lead exposure, *Brain Res.* 790, 98–107.

Guilarte, T.R., McGlothan, J.L., and Nihei, M.K. (2000), Hippocampal expression of *N*-methyl-D-aspartate receptor (NMDAR1) subunit splice variant mRNA is altered by developmental exposure to Pb^{2+}, *Mol. Brain Res.* 76, 299–306.

Guilarte, T.R. et al. (2003), Environmental enrichment reverses cognitive and molecular deficits induced by developmental lead exposure, *Ann. Neurol.* 53, 50–56.

Gutoswski, M. et al. (1998), Synaptic plasticity in the CA1 and CA3 hippocampal region of pre-and postnatally lead-exposed rats, *Toxicol. Lett.* 95, 195–203.

Harris, K.H., and Teyler, T.J. (1984), Developmental onset of long-term potentiation in area CA1 of the rat hippocampus, *J. Physiol.* 346, 27–48.

Hashemzadeh-Gargari, H., and Guilarte, T.R. (1999), Divalent cations modulate *N*-methyl-D-aspartate receptor function at the glycine site, *J. Pharm. Exp. Ther.* 290, 1356–1362.

Hattori, H., and Wasterlain, C.G. (1990), Excitatory amino acids in the developing brain: ontogeny, plasticity and excitotoxicity, *Pediatr. Neurol.* 6, 219–228.

Hebb, D.O. (1949), *The Organization of Behavior*, Wiley, New York.

Hollmann, M., and Heinemann, S. (1994), Cloned glutamate receptors, *Annu. Rev. Neurosci.* 17, 31–108.

Hoffmann, H., Hatt, H., and Gottmann, K. (1997), Presynaptic exocytosis regulates NR2A mRNA expression in cultured neocortical neurons, *NeuroReport* 8, 3731–3735.

Hoffmann, H. et al. (2000), Synaptic-activity-dependent developmental regulation of NMDA receptor subunit expression in cultured neocortical neurons, *J. Neurochem.* 75, 1590–1599.

Honer, M. et al. (1998), Differentiation of glycine antagonist sites of N-methyl-D-aspartate receptor subtypes, *J. Biol. Chem.* 273, 11158–11163.

Izquierdo, I. (1993), Long-term potentiation and the mechanism of memory, *Drug Dev. Res.* 30, 1–17.

Jett, D. A. et al. (1997), Age-dependent effects of developmental lead exposure on performance in the Morris water maze, *Pharmacol. Biochem. Behav.* 57, 271–279.

Kaul, B. et al. (1999), Follow-up screening of lead-poisoned children near an auto battery recycling plant, Haina, Dominican Republic, *Environ. Health Perspect.* 107, 917–920.

Kennedy, M.B. (1998), Signal transduction molecules at the glutamatergic postsynaptic membrane, *Brain Res. Rev.* 26, 243–257.

Kentros, C. et al. (1998), Abolition of long-term stability of new hippocampal place cells maps by NMDA receptor blockade, *Science* 280, 2121–2126.

Kew, J.N.C. et al. (1998), Developmental changes in NMDA receptor glycine affinity and ifenprodil sensitivity reveal three distinct populations of NMDA receptors in individual cortical neurons, *J. Neurosci.* 18, 1935–1943.

Kiyama, Y. et al. (1998), Increased threshold for long-term potentiation and contextual learning in mice lacking the NMDA-type glutamate receptors ε1 subunit, *J. Neurosci.* 18, 6704–6712.

Kornau, H-C. et al. (1995), Domain interaction between NMDA receptor subunits and the postsynaptic density protein PSD-95, *Science* 269, 1737–1740.

Kuhlmann, A. C., McGlothan, J. L., and Guilarte, T. R. (1997), Developmental lead exposure causes spatial learning deficits in adult rats, *Neurosci. Lett.* 233, 101–104.

Landrigan, P.J. et al. (2002), Environmental pollutants and disease in American children: estimates of morbidity, mortality, and costs for lead poisoning, asthma, cancer, and developmental disabilities, *Environ. Health Perspect.* 110, 721–728.

Lanphear, B.P. et al. (1999), Primary prevention of childhood lead exposure: a randomized trial of dust control, *Pediatrics* 103, 772–777.

Lanphear, B.P. et al. (2000), Cognitive deficits associated with blood lead concentrations < 10 µg/dL in US children and adolescent, *Public Health Rep.* 115, 521–529.

Lasley, S.M., and Gilbert, M.E. (1996), Presynaptic glutamatergic function in dentate gyrus *in vivo* is diminished by chronic exposure to inorganic lead, *Brain Res.* 736, 125–134.

Lasley, S.M., Green, M.C., and Gilbert, M.E. (2001), Rat hippocampal NMDA receptor binding as a function of chronic lead exposure level, *Neurotoxicol. Teratol.* 23, 185–189.

Lasley, S.M., Polan-Curtain, J., and Armstrong, D.L. (1993), Chronic exposure to environmental levels of lead impairs *in vivo* induction of long-term potentiation in rat hippocampal dentate, *Brain Res.* 614, 347–351.

Laube, B., Kuhse, J., and Betz, H. (1998), Evidence for a tetrameric structure of recombinant NMDA receptors, *J. Neurosci.* 18, 2954–2961.

Laurie, D.J., and Seeburg, P.H. (1994), Regional and developmental heterogeneity in splicing of the rat brain NMDAR1 mRNA, *J. Neurosci.* 14, 3180–3194.

Lin, J.W. et al. (1998), Yotiao, a novel protein of neuromuscular junction and brain that interacts with specific splice variants of NMDA receptor subunit NR1, *J. Neurosci.* 18, 2017–2027.

Lindlbauer, R. et al. (1998), Regulation of kinetic and pharmacological properties of synaptic NMDA receptors depends on presynaptic exocytosis in rat hippocampal neurones, *J. Physiol.* 508, 495–502.

Lopez-Carillo, L. et al. (1996), Prevalence and determinants of lead intoxication in Mexican children of low socioeconomic status, *Environ. Health Perspect.* 104, 1208–1211.

Luo, J. et al. (1996), Ontogeny of NMDA NR1 subunit protein expression in five regions of rat brain, *Dev. Brain Res.* 92, 10–17.

Madison, D.V., Malenka, R.C., and Nicoll, R.A. (1991), Mechanisms underlying long-term potentiation of synaptic transmission, *Annu. Rev. Neurosci.* 14, 379–397.

Malenka, R.C., and Nicoll, R.A. (1993), NMDA-receptor-dependent synaptic plasticity: multiple forms and mechanisms, *TINS* 16, 521–527.

Malenka, R.C., and Nicoll, R.A. (1999), Long-term potentiation—A decade of progress? *Science* 285, 1870–1874.

McIlhinney, R.A.J. et al. (1996), Cell surface expression of the human *N*-methyl-D-aspartate receptor subunit 1a requires the co-expression of the NR2A subunit in transfected cells, *Neuroscience* 70, 989–997.

McIlhinney, R.A.J. et al. (1998), Assembly intracellular targeting and cell surface expression of the human *N*-methyl-D-aspartate receptor subunits NR1a and NR2A in transfected cells. *Neuropharmacology* 37, 1355–13667.

McMichael, A.J. et al. (1988), Port Pirie cohort study: environmental exposure to lead and children's abilities at the age of four years, *New Engl. J. Med.* 319, 468–475.

McNamara, R.K., and Skelton, R.W. (1993), The neuropharmacological and neurochemical basis of place learning in the Morris water maze, *Brain Res. Rev.* 18, 33–49.

Mielke, H.W. (1999), Lead in the inner cities. *Am. Scientist* 87, 62–73.

Monyer, H. et al. (1992), Heteromeric NMDA receptors: molecular and functional distinction of subtypes, *Science* 256, 1217–1221.

Monyer, H. et al. (1994), Developmental and regional expression in the rat brain and functional properties of four NMDA receptors, *Neuron* 12, 529–540.

Moriyoshi, K. et al. (1991), Molecular cloning and characterization of the rat NMDA receptor, *Nature* 354, 31–37.

Morris, R.G.M. et al. (1986), Selective impairment of learning and blockade of long-term potentiation by an *N*-methyl-D-aspartate receptor antagonist, AP5, *Nature* 319, 774–776.

Nakazawa, K. et al. (2002), Requirement for hippocampal CA3 NMDA receptors in associative memory recall, *Science* 297, 211–218.

Needleman, H.L. (1998), Childhood lead poisoning: the promise and abandonment of primary prevention, *Am. J. Public Health* 88, 1871–1877.

Nihei, M.K., and Guilarte, T.R. (1999), NMDAR-2A subunit protein expression is reduced in the hippocampus of rat exposed to Pb^{2+} during development, *Mol. Brain Res.* 66, 42–49.

Nihei, M. K. et al. (2000), *N*-methyl-D-aspartate receptor subunit changes are associated with lead-induced deficits of long-term potentiation and spatial learning, *Neuroscience* 99, 233–242.

Nihei, M. K. and Guilarte, T.R. (2001), Molecular changes in glutamatergic synapses induced by Pb^{2+}: Association with deficits of LTP and spatial learning, *NeuroToxicology* 22, 635–643.

Nilsson, M. et al. (1999), Enriched environment increases neurogenesis in the adult rat dentate gyrus and improves spatial learning, *J. Neurobiol.* 39, 569–578.

Nowak, L. et al. (1984), Magnesium gates glutamate-activated channels in mouse central neurons, *Nature* 307, 462–465.

O'Brien, R.J., Lau, L-F., and Huganir, R.L. (1998), Molecular mechanisms of glutamate receptor clustering at excitatory synapses, *Curr. Opin. Neurobiol.* 8, 364–369.

Okabe, S., Miwa, A., and Okado, H. (1999), Alternative splicing of the C-terminal domain regulates cell surface expression of the NMDA receptor NR1 subunit, *J. Neurosci.* 19, 7781–7792.

Ozawa, S., Kamiya, H., and Tsuzuki, K. (1998), Glutamate receptors in the mammalian central nervous system, *Prog. Neurobiol.* 54, 581–618.

Parada-Turska, J., and Turski, W.A. (1990), Excitatory amino acid antagonists and memory: effects of drugs acting at *N*-methyl-D-aspartate receptors in learning and memory tasks, *Neuropharmacology* 29, 1111–1116.

Pirkle, J.L. et al. (1994), The decline in blood lead levels in the United States: the National Health and Nutrition Examination Surveys (NHANES), *J. Am. Med. Assoc.* 272, 284–291.

Pirkle, J.L. et al. (1998), Exposure of the U.S. population to lead, 1991–1994, *Environ. Health Perspect.* 106, 745–750.

Quinlan, E.M. et al. (1999a), Rapid, experience-dependent expression of synaptic NMDA receptors in visual cortex *in vivo*, *Nature Neurosci.* 2, 352–357.

Quinlan, E.M., Olstein, D.H., and Bear, M.F. (1999b), Bidirectional, experience-dependent regulation of *N*-methyl-D-aspartate receptor subunit composition in the rat visual cortex during development, *Proc. Natl. Acad. Sci.* 96, 12876–12880.

Rice, D. C. (1984), Behavioral deficit (delayed matching to sample) in monkeys exposed from birth to low levels of lead, *Toxicol. Appl. Pharmacol.* 75, 337–345.

Rice, D. C. (1990), Lead-induced behavioral impairment on a spatial discrimination reversal task in monkeys exposed during different periods of development, *Toxicol. Appl. Pharmacol.* 106, 327–333.

Riva, M.A. et al. (1994), Regulation of NMDA receptor subunit mRNA expression in the rat brain during postnatal development, *Mol. Brain Res.* 25, 209–216.

Romieu, I., Lacasana, M., McConnell, R., and the Lead Research Group of the Pan-American Health Organization (1997), Lead exposure in Latin America and the Caribbean, *Environ. Health Perspect.* 105, 398–405.

Ruan, D-Y. et al. (1998), Impairment of long-term potentiation and paired-pulse facilitation in rat hippocampal dentate gyrus following developmental lead exposure in vivo, *Brain Res.* 806, 196–201.

Rubin, C.H. et al. (2002), Lead poisoning among children in Russia: concurrent evaluation of childhood lead exposure in Ekaterinburg, Krasnouralsk, and Volgograd, *Environ. Health Perspect.* 110, 559–562.

Sakimura, K. et al. (1995), Reduced hippocampal LTP and spatial learning in mice lacking NMDA receptor ε1 subunit, *Nature* 373, 151–155.

Satcher, D.S. (2000), The Surgeon General on the continuing tragedy of childhood lead poisoning, *Public Health Rep.* 115, 579–580.

Scheetz, A.J., and Constantine-Paton, M. (1994), Modulation of NMDA receptor function: implications for vertebrate neural development, *FASEB J.* 8, 745–752.

Scott, D.B. et al. (2001), An NMDA receptor ER retention signal regulated by phosphorylation and alternative splicing, *J. Neurosci.* 21, 3063–3072.

Shimizu, E. et al. (2000), NMDA receptor-dependent synaptic reinforcement as a crucial process for memory consolidation, *Science* 290, 1170–1174.

Stocca, G., and Vicini, S. (1998), Increased contribution of NR2A subunit to synaptic NMDA receptors in developing rat cortical neurons, *J. Physiol.* 507, 13–24.

Sucher, N.J. et al. (1995), Developmental and regional expression pattern of a novel NMDA receptor-like subunit (NMDAR-L) in the rodent brain, *J. Neurosci.* 15, 6509–6520.

Sun, L. et al. (1998), Identification of a long variant of mRNA encoding the NR3 subunit of the NMDA receptor: its regional and developmental expression in the rat brain, *FEBS Lett.* 441, 392–396.

Teyler, T.J., and DiScenna, P. (1987), Long-term potentiation, *Annu. Rev. Neurosci.* 10, 131–161.

Thomas, K.L. et al. (1994), Regulation of the expression of NR1 NMDA glutamate receptor subunits during hippocampal LTP, *NeuroReport* 6, 119–123.

Tsien, J.Z., Huerta, P.T., and Tonegawa, S. (1996), The essential role of hippocampal CA1 NMDA receptor-dependent synaptic plasticity in spatial learning, *Cell* 87, 1327–1338.

Tsumoto, T. (1992), Long-term potentiation and long-term depression in the neocortex, *Prog. Neurobiol.* 39, 209–228.

Van Praag, H., Kempermann, G., and Gage, F.H. (2000), Neural consequences of environmental enrichment, *Nat. Rev. Neurosci.* 1, 191–198.

Vicini, S. et al. (1998), Functional and pharmacological differences between recombinant N-methyl-D-aspartate receptors, *J. Neurophysiol.* 79, 555–566.

Walker, D.L., and Gold, P.E. (1991), Effects of a novel NMDA antagonist, NPC 12626, on long-term potentiation, learning and memory, *Brain Res.* 549, 213–221.

Wechsler, A., and Teichberg, V. (1998), Brain spectrin binding to the NMDA receptor is regulated by phosphorylation, calcium and calmodulin, *EMBO J.* 17, 3931–3939.

White, R.F. et al. (1993), Residual cognitive deficits 50 years after lead poisoning during childhood, *Br. J. Ind. Med.* 50, 613–622.

Williams, K. (1993), Ifenprodil discriminates subtypes of the N-methyl-D-aspartate receptor: selectivity and mechanisms at recombinant heteromeric receptors, *Mol. Pharmacol.* 44, 851–859.

Williams, K. et al. (1993), Developmental switch in the expression of NMDA receptors occurs *in vivo* and *in vitro*, *Neuron* 10, 267–278.

Winneke, G., Brockhaus, A., and Baltissen, R. (1977), Neurobehavioral and systemic effects of longterm blood lead elevation in rats, *Arch. Toxicol.* 37, 247–263.

Winneke, G. et al. (1994), Neurobehavioral and neurophysiological observations in six year old children with low levels of lead in East and West Germany, *NeuroToxicology* 15, 705–714.

Wyszynski, M. et al. (1998), Differential regional expression and ultrastructural localization of alpha-actinin-2, a putative NMDA receptor-anchoring protein, in rat brain, *J. Neurosci.* 18, 1383–1392.

Zaiser, A.E., and Miletic, V. (1997), Prenatal and postnatal chronic exposure to low levels of inorganic lead attenuates long-term potentiation in the adult rat hippocampus in vivo, *Neurosci. Lett.* 239, 128–130.

Zalusky, R.A., and Nicoll, R.A. (1990), Comparison of two forms of long-term potentiation in single hippocampal neurons, *Science* 248, 1619–1624.

Zhang, X-Y. et al. (2002), Effect of developmental lead exposure on the expression of specific NMDA receptor subunit mRNAs in the hippocampus of neonatal rats by digoxigenin-labeled *in situ* hybridization histochemistry, *Neurotoxicol. Teratol.* 24, 149–160.

Zhong, J. et al. (1995), Expression of mRNAs encoding subunits of the NMDA receptor in developing rat brain, *J. Neurochem.* 64, 531–539.

Ziff, E.B. (1997), Enlightening the postsynaptic density, *Neuron* 19, 1163–1174.

Zukin, R.S., and Bennett, M.V.L. (1995), Alternatively spliced isoforms of the NMDAR1 receptor subunit, *TINS* 18, 305–313.

CHAPTER **6**

Genetic and Toxicological Models of Neurodegenerative Diseases: Stepping Stones Toward Finding a Cure

Arezoo Campbell and Stephen C. Bondy

CONTENTS

6.1 INTRODUCTION

The etiology of idiopathic forms of neurodegenerative disorders is multifactorial. Combinations of environmental and genetic factors as well as the aging process contribute to the pathology involved. While the possible interconnection of such events makes it difficult to distinguish the single factors, the advances made in molecular biology techniques make it possible to at least identify how mutations in single genes alter homeostasis and lead to diseased states.

The first part of this chapter aims at describing transgenic animal models of specific neurodegenerative disorders. These molecular models further the understanding of how a particular gene and its products are involved in the promotion of neurodegeneration. Such models not only delineate the mechanism of pathogenesis, but can also provide a strong tool for development of pharmaceutical and nutritional supplementation, which may attenuate the rate of neurodegeneration.

The second part of this chapter focuses on environmental toxins that selectively affect the central nervous system. Epidemiological studies involving both identical and nonidentical twins suggest that environmental agents are important contributors to the onset and progression of neurodegenerative disorders. Xenobiotic factors may either play a direct role in initiation of neurodegeneration or can exacerbate underlying mechanisms in the disease process.

6.2 MOLECULAR MODELS OF NEURODEGENERATIVE DISORDERS

The two hallmarks of Alzheimer's disease (AD) are neurofibrillary tangles (NFT) and senile plaques (SP). The former are composed of paired helical filaments (PHFs) made of hyperphosphorylated tau proteins, and the latter are a core of aggregated amyloid-β (Aβ) peptides surrounded by reactive glia. Epidemiological studies have shed some light on the pathogenesis of familial forms of AD and it has been demonstrated that abnormalities in specific genes are responsible. These in turn have led to the development of transgenic animals overexpressing mutated forms of such disease-associated genes that help unravel the possible mechanisms underlying disease progression.

In order to study amyloid deposition, which leads to eventual senile plaque formation, several lines of amyloid precursor protein (APP) over mutated mice have been designed. One such model is the APP23 mouse that expresses a mutant form of the human APP. As they age, these animals develop amyloid plaques predominantly in the hippocampus and neocortex (Bondolfi et al., 2002). β-site-APP cleaving enzyme (BACE) has been shown to be involved in the processing of APP (Vassar et al., 1999). In a double transgenic mouse model, it has been demonstrated that overexpression of BACE in the APP23 animals leads to a robust cleaving of APP. This in turn causes an increase in the formation of Aβ1-42, the more amyloidogenic form of the protein, in the brain of these mice (Bodendorf et al., 2002). It is possible that such accelerated Aβ formation can lead to earlier deposition of amyloid plaques.

While APP mutation leads to Aβ production and plaque formation, transgenic animals expressing mutant human tau protein form NFT. In a mouse transgenic model overexpressing a mutated form of the human tau protein isolated from frontotemporal dementia parkinsonism brains, aggregated forms of the protein were found in hippocampal neurons. These irregular-looking cells contained NFT similar to those which characterize AD. Abnormal histology was accompanied by decrease in neural activity and behavioral impairments (Tanemura et al., 2002). However, mice expressing normal human tau accumulated phosphorylated tau proteins but did not form NFT, even in the presence of mutated presenilin-1 (Boutajangout et al., 2002). Thus, it appears that overexpression of the mutated form of the tau protein is essential for NFT formation in transgenic animals.

While the pathology of AD consists of both senile plaques and neurofibrillary tangles, transgenic animal models have largely been selected for either amyloid deposition or NFT formation. The brain of double transgenic animals that express both mutated APP and tau proteins show a greater similarity to the AD brain in that NFT and amyloid plaques as well as neuronal loss are all present (Lewis et al., 2001).

While AD is associated with accumulation of Aβ protein, Parkinson's disease (PD) is associated with the accumulation of alpha synuclein (α-syn). This is an evolutionarily well-conserved phosphoprotein expressed in the brain and believed to play an important role in synaptic plasticity. Mutations in the α-syn gene, lead to aggregation of the protein and Lewy body formation, and have been linked to relatively uncommon genetic variants of Parkinson's disease (Kahle et al., 2000). In primary human embryonic cells derived from the mesencephalon, it has been demonstrated that dopaminergic neurons are selectively vulnerable to α-syn overexpression (Zhou et al., 2002).

In an effort to understand the mechanism by which dopaminergic cells are more sensitive to the mutant form of the protein, Tabrizi and colleagues (2000a) created stable human-derived cells expressing either wild-type or mutant α-syn. They found that in the latter case, dopamine was much more toxic than in the cell lines expressing wild-type α-syn. This suggests that the mutation might make the cells more sensitive to dopamine toxicity and consequent degeneration. Recently, a group of scientists have developed a transgenic mouse expressing an alpha synuclein mutation that leads to adult-onset development of disease which closely mimics PD symptoms (Lee et al., 2002). These transgenic animals may provide new insights into the cause and possible treatment of the disorder.

While aggregation of Aβ is a component of AD pathology, and accumulation of α-synuclein is associated with PD, there are patients exhibiting both features. In a double transgenic mouse model overexpressing both APP and α-syn, the animals developed impairments in learning and memory and there was an earlier onset of motor deficits compared to animals only overexpressing alpha synuclein. In isolated cell lines, Aβ causes aggregation of α-synuclein and its intracellular accumulation (Masliah et al., 2001). Thus, agents that block over-production of Aβ may also be useful for treatment of other neurodegenerative disorders such as PD.

Amyotrophic Lateral Sclerosis (ALS) is a progressive neurodegenerative disease in which motor neurons in the brain and spinal cord are selectively destroyed. Approximately 98% of the cases are idiopathic. Familial forms have been linked to a mutation in the copper-zinc superoxide dismutase (SOD1) gene. A transgenic model for the disease, expressing mutant human SOD1, has been established. These mice are referred to as G93A and are used to study the pathology of the disease and possible cures. Although the mechanism of neuronal death in ALS is unknown, the availability of these transgenic mice has lead to new findings showing that a multitude of events may be occurring in the pathogenesis of the disease.

The formation of advanced glycation end products is selectively increased in the astrocytes of both ALS patients with the SOD1 mutation and in the transgenic mouse model, while this parameter is unaltered in residual neuronal cells (Shibata et al., 2002). It has been questioned whether expression of SOD mutations in motor neurons can lead to their demise (Lino et al., 2002). While high levels of mutant protein accumulate in these cells, this does not lead to the pathological lesions found in motor neurons of ALS patients or the transgenic animals. Thus, it appears that neuronal death may not be a consequence of nonfunctional SOD aggregation.

However, neuronal nitric oxide synthase (nNOS), an enzyme that can effect peroxynitrite production and thus oxidative changes, increases steadily in spinal motor neurons of the G93A mice with disease progression. The subsequent increase in NO and thence peroxynitrite production may contribute to the demise of the motor neurons (Sasaki et al., 2002).

There is other evidence that motor neuron cell death associated with ALS is due to oxidative stress. In the transgenic G93A animals, mitochondrial respiration and ATP synthesis are impaired. Since most of the active SOD1 is localized in the intermembrane space of mitochondria, and the mutated enzyme may lose its antioxidant function, this may lead to reactive oxygen species (ROS) damage to organelles (Mattiazi et al., 2002). These defects in mitochondria and ATP synthesis are not observed in presymptomatic animals or those carrying the wild type human SOD1 gene. Therefore, the mitochondrial derangement effect appears to be a consequence of the disease process rather than an initiating factor. A causal relationship has, however, been established between oxidative damage and motor neuron degeneration. Mouse motor neuron-like cells, transfected with the mutant human SOD1, have increased levels of oxidative stress and mitochondrial dysfunction compared to cells transfected with the wild-type gene. This leads to cell death, which can be prevented by simultaneous overexpression of antioxidant genes or by treatment with the spin-trapping molecule, DMPO. This species not only inhibits cell death but delays paralysis and enhances survival in G93A transgenic mice (Liu et al., 2002).

6.3 ENVIRONMENTAL TOXINS AS MODELS OF NEURODEGENERATION

A relation between aluminum (Al) exposure and human health has been postulated. In several instances one can observe a direct causal role between exposure to the metal and abnormal neurological symptoms (Alfrey et al., 1976; Rifat et al., 1990; Russo et al., 1991; Bishop et al., 1997; Authier et al., 2001). Epidemiological studies report a good correlation between chronic exposure to Al in the drinking water and a higher incidence of AD (Armstrong et al., 1996; McLachlan et al., 1996; Rondeau et al., 2000). High levels of aluminum have been shown in postmortem brain samples from Alzheimer's patients (Good et al., 1992) but this issue is controversial (Bjertness et al., 1996). However, the metal has been shown to play a role in the formation of the pathological lesions characteristic of AD.

Perl and Brody (1980) originally described the presence of aluminum in neurofibrillary tangles. The frontal cortex of renal dialysis patients not displaying dialysis encephalopathy has been analyzed in order to determine whether or not exposure to high levels of aluminum causes changes in the tau protein similar to those seen in AD. In the white matter, there was a correlation between an increase in the amount of truncated tau protein and an increase in aluminum content. Although the gray matter did not show an increase in truncated tau, there was a decrease in the level of normal tau (Harrington et al., 1994). Circular dichroism and NMR spectroscopy studies reveal that aluminum does bind to tau and, by doing so, produces the aggregation of the protein (Madhav et al., 1996).

Aluminum also modulates the levels of Aβ. Both *in vivo* and *in vitro* analysis reported that aluminum did not effect the processing or expression of APP (Neill et al., 1996). However, a more recent study found that Al exacerbates oxidative stress, Aβ deposition, and plaque formation in the brain of transgenic mice that overexpress APP (Pratico et al., 2002). Aluminum has also been shown to cause aggregation of Aβ (Mantyh et al., 1993; Kawahara et al., 1994; Bondy and Truong, 1999).

Al exposure has been proposed to play a role in the progressive neurodegenerative disease found in Guam and neighboring islands, amyotrophic lateral sclerosis and Parkinsonism dementia complex (ALS/PDC), whose pathology includes deposition of neurofibrillary tangles which contain Al (Strong and Garruto, 1997). However, since the incidence of the disease has declined in the regions where the disease was most prevalent, while the concentration of the metal in the soil and drinking water remains constant, the link has weakened. In contrast, there is a strong link between the use of the cycad plant in food supplies and geographic incidence of the disease (Zhang et al., 1996).

The strongest epidemiological link between ALS/PDC and an environmental toxin comes from the correlation between consumption of the cycad seed and prevalence of the disease. Another interesting hypothesis is that the consumption of flying foxes by the Chamorro people may have resulted in ALS-PDC. These animals feed on the cycad seed and thus accumulate a high concentration of the neurotoxin. Since the number of flying foxes has decreased and the animals are no longer consumed, the prevalence of the disease has gone down (Cox and Sacks, 2002). The disease incidence was high among the Chamorros of the Western Pacific islands of Guam and Rota. The number of cases has substantially decreased since World War II, which brought about the Americanization of the diet of the Chamorro population (Spencer et al., 1987).

Dietary administration of the Cycas amino acid beta-*N*-methylamino-L-alanine (BMAA) fed to macaque monkeys led to symptoms resembling ALS/PDC (Spencer et al., 1987). However, it was later demonstrated that this component of the cycad plant is probably not the causative factor since much of it is removed during processing (Duncan et al., 1990) and the level of cycasin (Figure 6.1) is approximately ten times greater than the levels of BMAA (Kisby et al., 1992). Therefore, the actual toxin in the cycad plant, which is responsible for the disease, has as yet not been definitively identified but appears to be cycasin (Wilson et al., 2002).

Male mice fed a diet containing washed cycad flour developed progressive motor as well as cognitive impairment and neurodegeneration was extensive in the hippocampus, spinal cord, neocortex, substantia nigra, and the olfactory bulb (Wilson et al., 2002). The exact mechanism of cell death is unknown. One study has shown that cycasin can inhibit DNA repair. The ability to repair DNA is also reduced in the brains of ALS/PDC, ALS, and AD patients. Neuronal cell death may be a consequence of enhanced DNA damage (Kisby et al., 1999). Interestingly, the decrease in cycasin-induced DNA repair co-occurs with accumulation of phosphorylated tau (Esclaire et al., 1999).

The most important finding linking an exogenous toxin to PD occurred when it was discovered that a group of young individuals developed symptoms of the disease

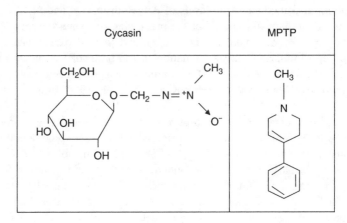

Figure 6.1   The chemical structure of cycasin and 1-methyl-4-phenyl-1,2,3,6-tetrahydro-pyridine (MPTP).

after using an impure form of synthesized heroin. The toxin was identified to be the activated form of 1-methyl-4-phenyl-1,2,3,6-tetrahydropyridine (MPTP) (Figure 6.1). Monoamine oxidase B was responsible for the transformation reaction of the compound to the 1-methyl-4-phenylpyridinium ion (MPP^+), which is transported to and concentrates within dopaminergic neurons through neurotransmitter uptake mechanisms. Thus, the toxic effect of MPP^+ is very specific to the nigrostriatal pathway, which is rich in dopaminergic cells.

It is important to assess if genetic mutations linked to PD may make an organism more prone to environmental insults leading to cell death. In one study it was demonstrated that following treatment with MPTP, the level of neuronal death in the substantia nigra was the same in a transgenic animal model expressing mutant alpha-synuclein, and wild-type control animals (Rathke-Hartlieb et al., 2001). Therefore, it appears that mutations in the α-syn, associated with familial PD, do not enhance susceptibility to MPTP. It may be that the genetic and nongenetic variants of PD, while presenting with the same progressive dopaminergic damage, are in fact initiated by unrelated processes.

6.4 COMMON MECHANISTIC LINKS BETWEEN NEURODEGENERATIVE DISEASES

One feature characterizing a wide range of disorders is enhanced oxidative damage. There is evidence of increased free radical injury in AD brain tissue (Marcus et al., 1998; Smith et al., 1998; Nunomura et al., 1999a). Whether this is a cause or the end result of the disease process is not certain. However, in Down's syndrome, which leads to early onset of AD, evidence of oxidative stress is found long before neuropathological changes (Busciglio and Yankner, 1995; Odetti et al., 1998; Nunomura et al., 2000). There is a significant increase in the level of free 4-hydroxynonenal (HNE), an end product of lipid peroxidation, in the amygdala, hippocampus and the hippocampal gyrus of AD patients compared to age-matched controls

(Markesbery and Lovell, 1997). When AD brains are treated with antibodies against 4-hydroxynonenal, neurofibrillary tangles, but also neurons lacking them, display elevated HNE-pyrrole immunoreactivity in comparison with age-matched controls (Sayre et al., 1997). Oxidative damage to macromolecules is thus not merely a consequence of a pre-existing deficit but may actually participate in the progression of the disease.

A recent study has looked at dietary components that may be related to AD and reported that low folic acid levels are linked to increased susceptibility to AD. It has been proposed that folic acid deficiency impairs DNA repair and thus potentiates neurodegeneration by enhancing Aβ-induced oxidative modification of DNA bases (Kruman et al., 2002). While aggregated forms of this peptide are known to generate free radicals, a new study shows that the monomeric Aβ1-40 actually functions as an antioxidant and, similar to vitamin E and glutathione, protects neurons against Fe(II)-induced oxidative damage (Zou et al., 2002). Therefore, it appears that the production of Aβ that occurs in normal brain may have a beneficial physiological role and only when the protein begins to form fibrils, it can then serve as a foci for increased and possibly detrimental ROS production.

The increasing evidence for oxidant events in AD is paralleled by findings that aluminum enhances cerebral pro-oxidant status. Aluminum treatment of experimental animals can cause increased cortical ROS production or lipid peroxidation (Ohtawa et al., 1993; Gupta and Shukla, 1995; Bondy et al., 1998a). Also, treatment of cell lines of CNS origin with the metal can promote ROS generation (Xie et al., 1996; Campbell et al., 1999).

In isolated cerebrocortical synaptosomes, the metal does not directly cause oxidative events but appears to markedly enhance the capacity of iron to produce ROS (Oteiza, 1994; Bondy and Kirstein, 1996). The mechanism of this promotion appears to be the stabilization of ferrous iron which is more potent in promoting the generation of ROS compared to ferric iron (Yang et al., 1999). This potentiation of oxidative events by Al is not limited to iron but can occur with other transition metals including chromium and copper (Bondy et al., 1998b). Both Al and Aβ stabilize iron in its ferrous form and by this means potentiates Fe-induced ROS production (Bondy et al., 1998c; Yang et al., 1999). The concept that the role of Al in the pathology of AD may be mediated by oxidative stress is supported by a recent study demonstrating that in transgenic mice overexpressing APP, Al enhances Aβ levels and plaque formation. This is prevented by co-exposure to vitamin E (Pratico et al., 2002).

The concept of oxidative stress as a common endpoint in different neurodegenerative diseases is strengthened by the observation that the levels of 8-hydroxy-2-deoxyguanosine (8-OHDG) and 8-hydroxyguanosine (8-OHG) are increased in patients with AD, PD, and ALS (Kikuchi et al., 2002). In PD, the lipid peroxidation end-product 4-HNE is immunolocalized with Lewy bodies and oxidative stress may play a role in their formation (Castellani et al., 2002). Coenzyme Q10, an important cofactor of the electron transport chain and an antioxidant (Beal, 2002), protects not only against MPTP toxicity in mice but also enhances survival in ALS transgenic animals. Thus, there is some evidence that oxidative stress may be a condition leading to subsequent neurodegeneration and antioxidants may ameliorate this effect.

Increased ROS formation in neuronal cells overexpressing normal human or mutated alpha synuclein genes (Junn and Mouradian, 2002) implies that this protein is involved in enhancing oxidative stress. This mirrors the pro-oxidant effect of $A\beta$ on neuronal cells and further strengthens the link between oxidative stress in different neurodegenerative disorders.

An age-related increase in inflammation within the brain has been repeatedly reported (David et al., 1997; Streit et al., 1999; Sharman et al., 2002). NF-κB is a transcription factor involved in the innate immune response (Baeuerle and Henkel, 1994; Medzhitov and Janeway, 1998; Hatada et al., 2000). There is a significant correlation between the amount of activated NF-κB and a key inflammatory enzyme, COX-2, in both the aging and AD brain (Lukiw and Bazan, 1998). The number of activated astrocytes is increased in AD brains and these are associated with senile plaques and cerebral microvessels (Cullen, 1997).

Cytokines are synthesized by activated glia in response to pathogens and trauma and both IL-1 as well as IL-6 levels are elevated in the AD brain (Cadman et al., 1994). Chronic production of these factors by activated glial cells may result in cytotoxicity because they recruit ROS producing macrophages which may lead to neuronal cell loss (Dunn, 1991). Systemic inflammation increases serum amyloid A (SAA) and in a transgenic mouse overexpressing the SAA1 gene, during an acute inflammatory response, amyloid A deposits appear in the brain. The nonsteroidal anti-inflammatory drug indomethacin inhibits this deposition (Guo et al., 2002). Inflammation can be exacerbated by exogenous stressors such as Al. The metal has been shown to promote $A\beta$ production and plaque deposition in APP transgenic mice (Pratico et al., 2002). Furthermore, intracerebroventricular injection of Al salts leads to activation of glial cells in the rat brain (Platt et al., 2001). Al caused an increase in TNFα production and NF-κB activation in a human glioblastoma cell line (Campbell et al., 2002) and mice exposed to Al in drinking water (Campbell et al., 2004).

Transgenic models of AD suggest that inflammatory events are involved in aging and pathology of the disease. Tg2576 mice show increases in cytokine mRNA levels at 16 months but not at 6 months (Sly et al., 2001). Microglial activation was not observable until mice were 12 months or older in an animal model expressing both mutated APP and presenilin-1 protein (PS1), although aggregated amyloid deposits were present around dystrophic neurites as well as activated astrocytes at an early age (Gordon et al., 2002). Therefore, it is possible that inflammatory processes do not necessarily initiate the pathological lesions in AD but may enhance the progression of the disease.

There is some evidence that controlled microglial activation may actually be advantageous to an aging brain. In APP and PS1 double transgenic mice, intracranial injection of lipopolysaccharide (LPS) in the hippocampus reduced $A\beta$ levels compared to animals which were only injected with saline (DiCarlo et al., 2001). Furthermore, two independent groups have demonstrated that in the ALS transgenic mouse model, treatment with glial derived neurotrophic factor (GDNF) has protective effects (Manabe et al., 2002; Acsadi et al., 2002). It appears that limited and confined inflammation may be beneficial. However, unresolved pathological lesions may lead to chronic activation of uncontrollable inflammatory events that can actually accentuate existing problems in the diseased brain.

Activated glial cells have been proposed to play a role in PD as well as AD. In a recent study it was demonstrated that the injection of LPS in the supranigral area of rat brains resulted in microglial and astroglial proliferation. These reactive glial cells express increased iNOS immunoreactivity and 3-nitrotyrosine formation leading to subsequent neuronal death (Iravani et al., 2002). The iNOS appears to be an important toxic mediator since pretreatment of animals with an iNOS inhibitor, substantially reduced the extent of cell death. Enhanced production of NO may potentiate already existing oxidative stress and this in turn may be the underlying cause of the decrease in neuronal viability.

It has been demonstrated that metamphetamine and MPTP-induced peroxynitrite formation may be a direct causative factor in dopaminergic neuron toxicity by nitrosylating the human dopamine transporter and thus rendering it ineffective (Park et al., 2002). Inhibition of glial cell activation in the mouse brain decreases MPTP-induced neurotoxicity by blocking the formation of iNOS and interleukin 1β (Wu et al., 2002), both of which are products of NF-κB-promoted gene transcription. In neuroblastoma cells, the MPP ion has been shown to activate NF-κB and the stress-activated c-Jun N-terminal kinase (Cassarino et al., 2000). However, another study looked at the effect of LPS induced toxicity in dopaminergic cells and showed that inhibition of iNOS did not enhance survival of these neurons. Instead, the levels of the cytokines TNF-α and IL-1β appeared to mediate some of the neurotoxicity observed (Gayle et al., 2002). Microglial activation is also believed to play a role in ALS-associated motor neuron death. An inhibitor of microglial activation, minocycline, has been shown not only to delay the onset of the disease but also to dose-dependently increase survival of the G93A transgenic mice (Van Den Bosch et al., 2002). Thus cell death that accompanies neurodegeneration and toxic insult may in part be mediated by chronic and uncontrollable glial activation.

6.5 CONCLUSION

The availability of selectively mutated transgenic animals has allowed the development of novel models of neurodegenerative diseases (Table 6.1). Certain environ-

Table 6.1   Example of Several Genes that Are Altered in Transgenic Animal Models Specific for the Indicated Neurodegenerative Diseases

Disease	Mutated or Overexpressed Genes
Alzheimer's disease (AD)	• Amyloid precursor protein (APP) • Beta-site-APP cleaving enzyme (BACE) • Presenilin-1 (PS1) • Tau
Parkinson's disease (PD)	• Alpha synuclein (α-syn)
Amyotrophic lateral sclerosis (ALS)	• Copper-zinc superoxide dismutase (SOD1)

Table 6.2 Environmental Agents Linked to Specific Neurodegenerative Disorders

Environmental Factors	Specific Neurodegenerative Disease
Aluminum (Al)	• AD • ALS/PDC
Cycasin	• ALS/PDC
1-methyl-4-phenyl-1,2,3,6-tetrahydropyridine (MPTP)	• PD

mental toxins have also been shown to have utility by mimicking the distinctive pathophysiology of these disorders (Table 6.2). Both genetic and chemically induced models, using experimental animals, have been developed for several other neurological diseases including Huntington's disease (HD). Many of the features of HD can be reproduced in the R6/1 transgenic mouse leading to an abnormally expanded CAG repeat within the ITI5 gene. Amplification of this gene sequence leads to elongated polyglutamine chains on the widely expressed 349-kd protein, huntingtin (Tabrizi et al., 2000b). In a parallel manner much of the histopathology of HD can be reproduced in a rodent model after systemic treatment with 3-nitropropionic acid, 3-NP (Borlongan et al., 1995). 3-NP is a mitochondrial toxin, specifically inhibiting succinic dehydrogenase, and children exposed to this neurotoxin following consumption of mildewed sugar cane in China have shown neurological impairment (Ludolph et al., 1991). The relation between effecting HD-like changes by genotypic alterations on a specific gene, or by use of an inhibitor of an enzyme common to all cells is unclear. However, both models of HD as well as the parent disease show evidence of oxidative damage in affected striatal regions, and drugs that cause amelioration of behavioral and chemical deficits of the 3-NP model are also effective in treating the mouse mutant model (Keene et al., 2002). The symptomatology of HD in these model systems may be initiated by several unrelated means, and this opens the possibility of study of interplay between genotypic and environmental factors.

Using both genetic and chemical tools, new insights into the mechanistic basis of neurodegeneration have been gained. Two consistent findings are that both oxidative stress and inflammatory events play an important role. These common findings suggest that the combination of a multifaceted dietary regimen together with pharmaceutical intervention may delay disease processes and this dual approach may be applicable to many forms of neurodegenerative disorders.

ACKNOWLEDGMENTS

This study was supported in part by grants from the National Institutes of Health (ES 7992 and AG-16794).

REFERENCES

Acsadi, G. et al. (2002), Increased survival and function of SOD1 mice after glial cell-derived neurotrophic factor gene therapy, *Hum. Gene Ther.* 13, 1047–1059.

Alfrey, A. C., leGendre, G. R., and Kaehny, W. D. (1976), The dialysis encephalopathy syndrome. Possible aluminum intoxication, *New Engl. J. Med.* 294, 184–188.

Armstrong, R. A., Winsper, S. J., and Blair, J. A. (1996), Aluminium and Alzheimer's disease, review of possible pathogenic mechanisms, *Dementia* 7, 1–9.

Authier, F. J. et al. (2001), Central nervous system disease in patients with macrophagic myofasciitis, *Brain* 124, 974–983.

Baeuerle, P. A., and Henkel, T. (1994), Function and activation of NF-κB in the immune system, *Annu. Rev. Immunol.* 12, 141–179.

Beal, M. F. (2002), Coenzyme Q10 as a possible treatment for neurodegenerative diseases, *Free Rad. Res.* 36, 455–460.

Bishop, N. J. et al. (1997), Aluminum neurotoxicity in preterm infants receiving intravenous-feeding solutions, *New Engl. J. Med.* 336, 1557–1561.

Bjerntness, E. et al. (1996), Content of brain aluminum is not altered in Alzheimer disease, *Alz. Dis. Assoc. Disorders* 10, 171–174.

Bodendorf, U. et al. (2002), Expression of the human beta-secretase in the mouse brain increases the steady-state level of beta-amyloid, *J. Neurochem.* 80, 799–806.

Bondolfi, L. et al. (2002), Amyloid-associated neuron loss and gliogenesis in the neocortex of amyloid precursor protein transgenic mice, *J. Neurosci.* 22, 515–522.

Bondy, S. C., and Kirstein, S. (1996), The promotion of iron-induced generation of reactive oxygen species in nerve tissue by aluminum, *Mol. Chem. Neuropathol.* 27, 185–194.

Bondy, S. C., Ali, S. F., and Guo-Ross, S. X. (1998a), Aluminum but not iron treatment induces pro-oxidant events in the rat brain, *Mol. Chem. Neuropathol.* 34, 219–232.

Bondy, S. C., Guo-Ross, S. X., and Pien, J. (1998b), Mechanisms underlying aluminum-induced potentiation of oxidant properties of transition metals, *NeuroToxicology* 19, 65–72.

Bondy, S. C., Guo-Ross, S., and Truong, A. T. (1998c), Promotion of transition metal-induced reactive oxygen species formation by β-amyloid, *Neurochem. Int.* 33, 51–54.

Bondy, S. C., and Truong A. (1999), Potentiation of beta-folding of β-amyloid peptide 25-35 by aluminum salts, *Neurosci. Lett.* 267, 25–28.

Borlongan, C. V. et al. (1995), Systemic 3-nitropropionic acid: behavioral deficits and striatal damage in adult rats, *Brain Res. Bull.* 36, 549–56.

Boutajangout, A. et al. (2002), Increased tau phosphorylation but absence of formation of neurofibrillary tangles in mice double transgenic for human tau and Alzheimer mutant (M146L) presenilin-1, *Neurosci. Lett.* 18, 29–33.

Busciglio, J., and Yankner, B. A. (1995), Apoptosis and increased generation of reactive oxygen species in Down's syndrome neurons in vitro, *Nature* 378, 776–779.

Cadman, E. D., Witte, D. G., and Lee, C. M. (1994), Regulation of the release of interleukin-6 from human astrocytoma cells, *J. Neurochem.* 63, 980–987.

Campbell, A., Prasad, K. N., and Bondy, S. C. (1999), Aluminum induced oxidative events in cell lines: glioma are more responsive than neuroblastoma, *Free Rad. Biol. Med.* 26, 1166–1171.

Campbell, A. et al. (2002), Pro-inflammatory effects of aluminum in human glioblastoma cells, *Brain Res.* 933, 60–65.

Campbell, A. et al. (2004), Chronic Exposure to Aluminum in Drinking Water Increases Inflammatory Parameters Selectively in the Brain, *J. Neurosci. Res.* (In Press).

Cassarino, D. S. et al. (2000), Interaction among mitochondria, mitogen-activated protein kinase, and nuclear factor-kappaB in cellular models of Parkinson's disease, *J. Neurochem.* 74, 1384–1392.

Castellani, R. J. et al. (2002), Hydroxynonenal adducts indicate a role for lipid peroxidation in neocortical and brainstem Lewy bodies in humans, *Neurosci. Lett.* 319, 25–28.

Cox, P. A., and Sacks, O. W. (2002), Cycad neurotoxins, consumption of flying foxes, and ALS-PDC disease in Guam, *Neurology* 26, 956–959.

Cullen, K. M. (1997), Perivascular astrocytes within Alzheimer's disease plaques, *Neuroreport* 8, 1961–1966.

David, J. P. et al. (1997), Glial reaction in the hippocampal formation is highly concentrated with aging in human brain, *Neurosci. Lett.* 235, 53–56.

DiCarlo, G. et al. (2001), Intrahippocampal LPS injections reduce Abeta load in APP+PS1 transgenic mice, *Neurobiol. Aging* 22, 1007–1012:

Duncan, M. W. et al. (1990), 2-Amino-3-(methylamino)-propanoic acid (BMAA) in cycad flour: An unlikely cause of amyotrophic lateral sclerosis and parkinsonism-dementia of Guam, *Neurology* 40, 767–772.

Dunn, C. J. (1991), Cytokines as mediators of chronic inflammatory disease, *Cytokines and Inflammation*, ed. E. S. Kimball, CRC Press, Boca Raton, FL, 1–35.

Esclaire, F. et al. (1999), The Guam cycad toxin methylazoxymethanol damages neuronal DNA and modulates tau mRNA expression and excitotoxicity, *Exp. Neurol.* 155, 11–21.

Gayle, D. A. et al. (2002), Lipopolysaccharide (LPS)-induced dopamine cell loss in culture: roles of tumor necrosis factor-alpha, interleukin-1beta, and nitric oxide, *Brain Res. Dev. Brain Res.* 133, 27–35.

Good, P. F., Perl, D. P., Bierer, L., and Schmeidler J. (1992), Selective accumulation of aluminum and iron in the neurofibrillary tangles of Alzheimer's disease, a laser microprobe (LAMMA) study, *Ann. Neurol.* 31, 286–292.

Gordon, M. N. et al. (2002), Time course of the development of Alzheimer-like pathology in the double transgenic PS1+APP mouse, *Exp. Neurol.* 173, 183–195.

Guo, J. et al. (2002), Inflammation-dependent cerebral deposition of serum amyloid A protein in a mouse model of amyloidosis, *J. Neurosci.* 22, 5900–5909.

Gupta, A., and Shukla, G. S. (1995), Effect of chronic aluminum exposure on the levels of conjugated dienes and enzymatic antioxidants in hippocampus and whole brain of rat, *Bull. Environ. Contam. Toxicol.* 55, 716–722.

Harrington, C. R. et al. (1994), Alzheimer's-disease-like changes in tau protein processing: association with aluminum accumulation in brains of renal dialysis patients, *Lancet* 343, 993–997.

Hatada, E. N., Krappmann, D., and Scheidereit, C. (2000), NF-κB and the innate immune response, *Curr. Opin. Immunol.* 12, 52–58.

Iravani, M. M. et al. (2002), Involvement of inducible nitric oxide synthase in inflammation-induced dopaminergic neurodegeneration, *Neuroscience* 110, 49–58.

Junn, E., and Mouradian, M. M. (2002), Human alpha-synuclein overexpression increases intracellular reactive oxygen species levels and susceptibility to dopamine, *Neurosci. Lett.* 320, 146–150.

Kahle, P. J., Neumann, M., Ozmen, L., and Haass, C. (2000), Physiology and pathophysiology of alpha-synuclein. Cell culture and transgenic animal models based on a Parkinson's disease-associated protein, *Ann. NY Acad. Sci.* 920, 33–41.

Kawahara, M. et al. (1994), Aluminum promotes the aggregation of Alzheimer's amyloid β-protein in vitro, *Biochem. Biophys. Res. Commun.* 198, 531–535.

Keene, C. D. et al. (2002), Tauroursodeoxycholic acid, a bile acid, is neuroprotective in a transgenic animal model of Huntington's disease, *Proc. Natl. Acad. Sci. USA* 99, 10671–10676.

Kikuchi, A. et al. (2002), Systemic increase of oxidative nucleic acid damage in Parkinson's disease and multiple system atrophy, *Neurobiol. Dis.* 9, 244–248.

Kisby, G. E. et al. (1999), Damage and repair of nerve cell DNA in toxic stress, *Drug Metab. Rev.* 31, 589–618.

Kisby, G. E., Ellison, M., and Spencer, P. S. (1992), Content of the neurotoxins cycasin (methylazoxymethanol beta-D-glucoside) and BMAA (beta-N-methylamino-L-alanine) in cycad flour prepared by Guam Chamorros, *Neurology* 42, 1336–1340.

Kruman, I. I. et al. (2002), Folic acid deficiency and homocysteine impair DNA repair in hippocampal neurons and sensitize them to amyloid toxicity in experimental models of Alzheimer's Disease, *J. Neurosci.* 22, 1752–1762.

Lee, M. K. et al. (2002), Human alpha-synuclein-harboring familial Parkinson's disease-linked Ala-53 → Thr mutation causes neurodegeneration disease with alpha-synuclein aggregation in transgenic mice, *Proc. Natl. Acad. Sci. USA* 99, 8968–8973.

Lewis, J. et al. (2001), Enhanced neurofibrillary degeneration in transgenic mice expressing mutant tau and APP, *Science* 293, 1487–1491.

Lino, M. M., Schneider, C., and Caroni, P. (2002), Accumulation of SOD1 mutants in postnatal motoneurons does not cause motoneuron pathology or motoneuron disease, *J. Neurosci.* 22, 4825–4832.

Liu, R. et al. (2002), Increased mitochondrial antioxidative activity or decreased oxygen free radical propagation prevent mutant SOD1-mediated motor neuron cell death and increase amyotrophic lateral sclerosis-like transgenic mouse survival, *J. Neurochem.* 80, 488–500.

Ludolph, A. C. et al. (1991), 3-Nitropropionic acid-exogenous animal neurotoxin and possible human striatal toxin, *Can. J. Neurol. Sci.* 18, 492–498.

Lukiw, W. J., and Bazan, N. G. (1998), Strong nuclear factor-κB-DNA binding parallels cyclooxygenase-2 gene transcription in aging and in sporadic Alzheimer's disease superior temporal lobe neocortex, *J. Neurosci. Res.* 53, 583–592.

Madhav, T. R. et al. (1996), Preservation of native conformation during aluminum-induced aggregation of tau protein, *Neuroreport* 7, 1072–1076.

Manabe, Y. et al. (2002), Adenovirus-mediated gene transfer of glial cell line-derived neurotrophic factor prevents motor neuron loss of transgenic model mice for amyotrophic lateral sclerosis, *Apoptosis* 7, 329–334.

Mantyh, P. W. et al. (1993), Aluminum, iron and zinc ions promote aggregation of physiological concentrations of β-amyloid peptide, *J. Neurochem.* 61, 1171–1174.

Marcus, D. L. et al. (1998), Increased peroxidation and reduced enzyme activity in Alzheimer's disease, *Exp. Neurol.* 150, 40–44.

Markesbery, W. R. and Lovell, J. E. (1997), Four-hydroxynonenal, a product of lipid peroxidation, is increased in the brain in Alzheimer's disease, *Neurobiol. Aging* 19, 33–36.

Masliah, E. et al. (2001), Beta-amyloid peptides enhance alpha-synuclein accumulation and neuronal deficits in a transgenic mouse model linking Alzheimer's disease and Parkinson's disease, *Proc. Natl. Acad. Sci. USA* 98, 12245–12250.

Mattiazzi, M. et al. (2002), Mutated human SOD1 causes dysfunction of oxidative phosphorylation in mitochondria of transgenic mice, *J. Biol. Chem.* 16, 29626–29633.

McLachlan, D. R. C. et al. (1996), Risk for neuropathologically confirmed Alzheimer's disease and residual aluminum in municipal drinking water employing weighted residential histories, *Neurology* 46, 401–405.

Medzhitov, R., and Janeway, C. A. (1998), An ancient system of host defense, *Curr. Opin. Immunol.* 10, 12–15.

Neill, D. et al. (1996), Effect of aluminum on expression and processing of amyloid precursor protein, *J. Neurosci. Res.* 46, 395–403.

Nunomura, A. et al. (1999), RNA oxidation is a prominent feature of vulnerable neurons in Alzheimer's disease, *J. Neurosci.* 19, 1959–1964.

Nunomura, A. et al. (1999b), Neuronal RNA oxidation in Alzheimer's disease and Down's syndrome, *Ann. NY Acad. Sci.* 893, 362–364.

Odetti, P. et al. (1998), Early glycoxidation damage in brains from Down's syndrome, *Biochem. Biophys. Res. Commun.* 243, 849–851.

Ohtawa, M., Seko, M., and Takayama, F. (1993), Effect of aluminum ingestion on lipid peroxidation in rats, *Chem. Pharmacol. Bull.* 31, 1415–1418.

Oteiza, P. I. (1994), A mechanism for the stimulatory effect of aluminum on iron induced lipid peroxidation, *Arch. Biochem. Biophys.* 308, 374–379.

Park, S. U. et al. (2002), Peroxynitrite inactivates the human dopamine transporter by modification of cysteine 342: potential mechanism of neurotoxicity in dopamine neurons, *J. Neurosci.* 22, 4399–4405.

Perl, D. P., and Brody, A. R. (1980), Alzheimer's disease: x-ray spectrometric evidence of aluminum accumulation in neurofibrillary tangle-bearing neurons, *Science* 208, 297–299.

Platt, B. et al. (2001), Aluminum toxicity in the rat brain: histochemical and immunocytochemical evidence, *Brain Res. Bull.* 55, 257–267.

Pratico, D. et al. (2002), Aluminum modulates brain amyloidosis through oxidative stress in APP transgenic mice, *FASEB J.* 16, 1138–1140.

Rathke-Hartlieb, S. et al. (2001), Sensitivity to MPTP is not increased in Parkinson's disease-associated mutant alpha-synuclein transgenic mice, *J. Neurochem.* 77, 1181–1184.

Rifat, S. L. et al. (1990), Effect of exposure of miners to aluminum powder, *Lancet* 336, 1162–1165.

Rondeau, V. et al. (2000), Relation between aluminum concentrations in drinking water and Alzheimer's disease: an 8-year follow-up study, *Am. J. Epidemiol.* 152, 59–66.

Russo, L. S. et al. (1991), Aluminum toxicity in undialyzed patients with chronic renal failure, *J. Neurol. Neurosurg. Psychiat.* 55, 697–700.

Sasaki, S. et al. (2002), Neuronal nitric oxide synthase (nNOS) immunoreactivity in the spinal cord of transgenic mice with G93A mutant SOD1 gene, *Acta Neuropathol.* 103, 421–427.

Sayre, L. M. et al. (1997), 4-Hydroxynonenal-derived advanced lipid peroxidation end products are increased in Alzheimer's disease, *J. Neurochem.* 68, 2092–2097.

Sharman, K. G., Sharman, E., and Bondy, S. C. (2002), Dietary melatonin selectively reverses age-related changes in cortical basal cytokine mRNA levels, and their responses to an inflammatory stimulus, *Neurobiol. Aging* 3, 633–638.

Shibata, N. et al. (2002), Selective formation of certain advanced glycation end products in spinal cord astrocytes of humans and mice with superoxide dismutase-1 mutation, *Acta Neuropathol.* 104, 171–178.

Sly, L. M. et al. (2001), Endogenous brain cytokine mRNA and inflammatory responses to lipopolysaccharide are elevated in the Tg2576 transgenic mouse model of Alzheimer's disease, *Brain Res. Bull.* 56, 581–588.

Smith, M. A. et al. (1998), Amyloid-β deposition in Alzheimer transgenic mice is associated with oxidative stress, *J. Neurochem.* 70, 2212–2215.

Spencer, P. S. et al. (1987), Guam amyotrophic lateral sclerosis-parkinsonism-dementia linked to a plant extract neurotoxin, *Science* 237, 517–522.

Streit, W. J., Walter, S. A., and Pennell, N. A. (1999), Reactive microgliosis, *Prog. Neurobiol.* 57, 563–581.

Strong, M. J., and Garruto, R. M. (1997), Motor neuron disease, *Mineral and Metal Neurotoxicity*, ed. M. Yasui et al., CRC Press, Boca Raton, FL, 107.

Tabrizi, S. J. et al. (2000a), Expression of mutant alpha-synuclein causes increased susceptibility to dopamine toxicity, *Hum. Mol. Genet.* 9, 2683–2689.

Tabrizi, S. J. et al. (2000b), Mitochondrial dysfunction and free radical damage in the Huntington R6/2 transgenic mouse, *Ann. Neurol.* 47, 80–86.

Tanemura, K. et al. (2002), Neurodegeneration with tau accumulation in a transgenic mouse expressing V337M human tau, *J. Neurosci.* 22, 133–141.

Van Den Bosch, L. et al. (2002), Minocycline delays disease onset and mortality in a transgenic model of ALS, *Neuroreport* 13, 1067–1070.

Vassar, R. et al. (1999), Beta-secretase cleavage of Alzheimer's amyloid precursor protein by the transmembrane aspartic protease BACE, *Science* 286, 735–741.

Wilson, J. M. et al. (2002), Behavioral and neurological correlates of ALS-parkinsonism dementia complex in adult mice fed washed cycad flour, *Neuromol. Med.* 1, 207–221.

Wu, D. C. et al. (2002), Blockade of microglial activation is neuroprotective in the 1-methyl-4-phenyl-1,2,3,6-tetrahydropyridine mouse model of Parkinson disease, *J. Neurosci.* 22, 1763–1771.

Xie, C. X. et al. (1996), Intraneuronal aluminum potentiates iron-induced oxidative stress in cultured rat hippocampal neurons, *Brain Res.* 743, 271–277.

Yang, E. Y., Guo-Ross, S. X., and Bondy, S. C. (1999), The stabilization of ferrous iron by a toxic β-amyloid fragment and by an aluminum salt, *Brain Res.* 839, 221–226.

Zhang, Z. X. et al. (1996), Motor neuron disease on Guam: geographical incidence and familial occurrence, 1956–1985, *Acta Neurol. Scand.* 94, 51.

Zhou, W. et al. (2002), Overexpression of human alpha-synuclein causes dopamine neuron death in primary human mesencephalic culture, *Brain Res.* 926, 42–50.

Zou, K. et al. (2002), A novel function of monomeric amyloid β-protein serving as an antioxidant molecule against metal-induced oxidative damage, *J. Neurosci.* 22, 4833–4841.

CHAPTER 7

DNA Damage and DNA Repair in Neurotoxicology

Todd Stedeford and Fernando Cardozo-Pelaez

CONTENTS

7.1 GENERAL INTRODUCTION TO DNA DAMAGE AND REPAIR

The novice effects of most neurotoxins and neurotoxicants are dependent on their direct interaction or the structural modification of cellular macromolecules (e.g., lipids, proteins, or nucleic acids). A unique property of the central nervous system (CNS) is its composition of both actively dividing and differentiated cells, e.g., astrocytes and neurons, respectively. Both astrocytes and neurons are capable of *de novo* synthesis of proteins and lipids; however, unlike astrocytes, neurons do not replicate their genomic DNA and are more susceptible to toxicological outcomes following chemical exposures because of diminished levels of antioxidants, antioxidant enzymes, and DNA repair enzymes (Nouspikel and Hanawalt, 2000; Nouspikel and Hanawalt, 2002; Wang and Shum, 2002). Furthermore, the DNA in CNS is predisposed to a higher burden of damage from reactive oxygen species (ROS) because of CNS high rate of oxygen consumption and high metabolic requirements (Harman, 1993). Thus, an understanding of DNA damage and DNA repair in the context of the CNS is essential for elucidating processes of neurodegeneration

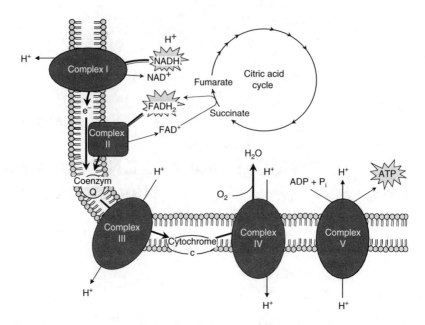

Figure 7.1   The respiratory enzyme complexes of the ETC transfer electrons (bold black line) between each subunit and ultimately to molecular oxygen (with the production of water). Protons are translocated across the inner mitochondrial membrane, which establishes a proton gradient and provides the energy for the production of ATP.

associated with idiopathic neurological deficits or those conditions resulting from exposure to neurotoxins or neurotoxicants.

The primary source of damage to macromolecules occurs from ROS that are generated during normal cellular respiration on the inner mitochondrial membrane from the electron transport chain (ETC). The citric acid cycle maintains the coenzymes NADH and flavoproteins in a reduced state and supplies reducing equivalents to the ETC (Figure 7.1). The ETC is comprised of five major enzyme complexes, including NADH CoQ reductase (complex I), succinate CoQ reductase (complex II), ubiquinol cytochrome c reductase (complex III), cytochrome c oxidase (COX; complex IV) and ATP synthetase (complex V).

The normal generation of ROS occurs at two main sites along the ETC, most of which are rapidly removed by the antioxidant systems in the cell. However, specific inhibitors of the ETC can cause a dramatic increase in ROS levels and override the antioxidant systems (Figure 7.2).

DNA is the target for a variety of endogenous and exogenous compounds that interact with the various structures of the double helix and ultimately result in modification of the genomic material. These modifications can be classified according to the major structural outcomes: single- and double-strand breaks, chemically modified bases, abasic sites, bulky adducts, inter- and intra-strand cross-linking, and base pair mismatches.

In general, the most common modifications of DNA occur as a consequence of unquenched ROS. Several types of ROS are known to exist, most of which are

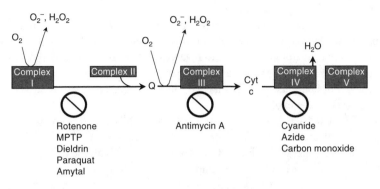

Figure 7.2   The major sites along the ETC where molecular oxygen (O_2) can escape. The resultant superoxide anion ($O_2^{\cdot-}$) can form hydrogen peroxide (H_2O_2) either spontaneously or enzymatically by manganese superoxide dismutase (MnSOD). The specific inhibitors listed have been shown to increase the production of ROS by blocking the flow of electrons at the respective sites.

generated during the reduction of molecular oxygen (O_2) to water on the ETC. In the cellular milieu, O_2 readily reacts to form partially reduced species that are short lived and highly reactive; these include superoxide anion ($O_2^{\cdot-}$), hydrogen peroxide (H_2O_2), and the most reactive species, the hydroxyl radical ($HO\cdot$) (Koppenol, 1990). It has been estimated that approximately 1–5% of the oxygen consumed by mitochondria is converted to ROS in the mitochondrial ETC (Sokoloff, 1960; Papa, 1996). $O_2^{\cdot-}$ is more reactive than H_2O_2, and although its reactivity with macromolecules is relatively low, H_2O_2 is capable of crossing cellular membranes. Therefore, H_2O_2 produced in the mitochondria could participate in radical reactions in the nucleus or some other organelle, rather than causing proximal damage as seen with $O_2^{\cdot-}$ or $HO\cdot$ (Reiter, 1995).

When levels of ROS surpass the protective mechanisms of a cell, they are capable of interacting with lipids, proteins, or nucleic acids (Moosmann and Behl, 2002). The outcomes of such interactions have been implicated in a variety of biological processes (Fridovich, 1978; Simic et al., 1989), not the least of which is aging (Harman, 1981). The occurrence of oxidative stress, caused by ROS, is believed to be a major etiological or pathogenic event of most degenerative processes. Exogenous agents such as pesticides, combustion products, ionizing radiation, etc. are known to generate radicals/oxidants. Importantly, it has been well documented that a variety of endogenous processes, in addition to cellular respiration, are significant generators of ROS, including dopamine metabolism, cytochrome P-450 detoxification reactions, phagocytic oxidative bursts, and peroxisomal leakage (Ames, 1983; Comporti, 1985; Fucci et al., 1983; Richter et al., 1988; Schraufstatter et al., 1988; Ames, 1989).

Because of the multitude of ROS that are continuously generated in the cell, numerous pathways have evolved that help to prevent or repair the damages that occur. Of these, antioxidants and antioxidant enzymes are the main pathways for preventing ROS-induced damage. In neural tissues, Cu/Zn-superoxide dismutase (Cu/Zn-SOD) and glutathione peroxidase (GPx) are the predominant oxygen scavenging enzymes (Marklund et al., 1983a; Jain et al., 1991a). However, these pathways are not absolute, and it is estimated that for every 10^{12} oxygen molecules entering a cell each day 1/100 damages lipids or proteins and 1/200 damages DNA. It is this damage to lipids, proteins, and DNA that makes the ROS so dangerous. Among the types of damage that can occur from ROS, damage to DNA has the most far-reaching implications for cell survival, and as such, countless redundancies in repair pathways have evolved.

The chemical basis for the effects of ROS on DNA is due to oxidation and chemical modification of the bases or the deoxyribose-phosphate backbone. When HO· radicals are generated adjacent to DNA, they attack both the deoxyribose sugar and the purine and pyrimidine bases and are capable of producing a wide range of products (Figure 7.3). Other oxygen species, such as singlet oxygen ($^1O_2$), preferentially attack guanine bases. $O_2^{·-}$ and H_2O_2 have a low reactivity with DNA; however, HO· radicals can be formed from either the $O_2^{·-}$ (Haber-Weiss reaction) or from H_2O_2 (Fenton reaction) (Breimer, 1991; Halliwell, 1994). Both reactions require a transitional metal such as iron or copper, and it is believed that the toxicity of $O_2^{·-}$ and H_2O_2 may be due to their conversion to the HO· radical. If such oxidized bases are transcribed and translated, they may lead to modifications in protein structure, and possibly an alteration in function, which could initiate a degenerative process (Rolig and McKinnon, 2000).

DNA repair pathways exist that are capable of recognizing thousands of altered bases. To counteract the deleterious effect of altered bases, a variety of DNA repair pathways have evolved in all organisms studied to date. The type of DNA damage determines the pathway deployed in the repair process. Helix distorting lesions like those formed by a variety of environmental agents, e.g., ionizing radiation, benzo[a]pyrene, or aflatoxin B_1, are repaired by the nucleotide excision repair (NER) pathway (Figure 7.4) (de Laat et al., 1999).

DNA damage addressed by the NER system is not equally repaired throughout the genome and two distinct subpathways have been identified: (1) transcription-coupled repair (TCR), the fast repair associated with actively transcribed genes; and (2) global genome repair (GGR), the slow repair associated with non-transcribed regions of the genome. Damage induced from alkylating agents, e.g., N-methyl-N-nitrosurea, is removed by O^6-alkylguanine-DNA alkyltransferase (MGMT), which covalently attaches to and removes the alkyl moiety, thus inactivating the enzyme (Figure 7.5) (de Laat et al., 1999).

The chemical modification of DNA bases by ROS or lipid peroxides may result in deamination (e.g., cytosine to uracil), alkylation (e.g., adenine to 3-methyladenine), oxidation (e.g., guanine to 8-oxyguanosine [oxo^{8}dG]), reduction (e.g., guanine to 2,6-diamino-4-hydroxy-5-formamidopyrimidine [fapy]), or errors in DNA replication (e.g., misincorporation of nucleotides). These types of structural modifications are addressed by enzymes of the base excision repair (BER) pathway (Figure 7.6) (Friedberg, 1995; de Laat et al., 1999). BER can proceed by two pathways designated as short-patch repair and long-patch repair (discussed below).

Quantitatively, lesions occurring from ROS are the most prominent type generated in the cell with an estimated 10,000 oxidative "hits" occurring per cell per day

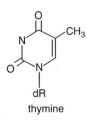

thymine

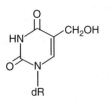

5-hydroxymethyluracil

thymine glycol

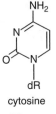

cytosine

(A)

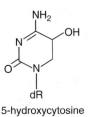

5-hydroxycytosine

5,6-dihydroxycytosine

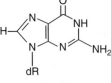

guanine

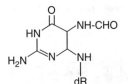

2,6-diamino-4-hydroxy-
5-formamidopyrimidine

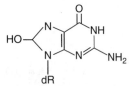

8-hydroxyguanine

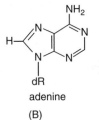

adenine

(B)

8-hydroxyadenine

2-hydroxyadenine

Figure 7.3 The hydroxyl radical is capable of generating a multitude of products from the normal bases of (A) pyrimidines and (B) purines. The normal base is shown on the left along side two representative bases that can be generated from hydroxyl radicals with the structural modifications shown in italics.

in the genome (Lindahl and Nyberg, 1972; Ames et al., 1993). The BER system is the most important pathway for removing these kinds of lesions and maintaining the integrity of the nuclear and mitochondrial genome; therefore, this pathway will be discussed in detail.

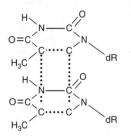

Cyclobutane pyrimidine dimer

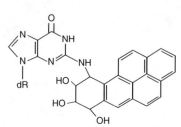

Adduct formation of BPDE with guanine

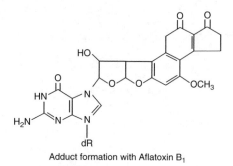

Adduct formation with Aflatoxin B$_1$

(A)

-DNA adduct recognition
-Oligonucleotide excision (~30 bp)
  by a multiprotein complex

-DNA polymerase
-DNA ligase
-Restoration of intact DNA

(B)

Figure 7.4 Typical adducts formed from environmental agents (A) ionizing radiation, benzo[a]pyrene, or aflatoxin B$_1$ are repaired by the (B) nucleotide excision repair pathway.

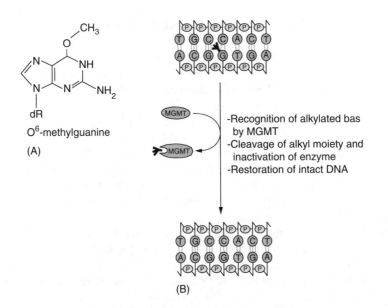

O^6-methylguanine

(A)

-Recognition of alkylated bas
 by MGMT
-Cleavage of alkyl moiety and
 inactivation of enzyme
-Restoration of intact DNA

(B)

Figure 7.5 Chemical modifications to DNA bases by alkylating agents (A) N-methyl-N-nitro-surea are removed by (B) alkylguanine-DNA alkyltransferase (MGMT).

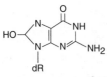

8-hydroxyguanine (oxo^{8}dG)

(A)

-Recognition/excision of
 oxidized base by N-glycosylase
-Cleavage of sugar-phosphate
 backbone

-DNA polymerase
-DNA ligase
-Restoration of intact DNA

(B)

Figure 7.6 Modification of the heterocyclic DNA bases that are caused by agents involved in normal cellular metabolism resulting in (A) oxidation (e.g., guanine to oxo^{8}dG) are corrected by enzymes of the (B) base excision repair (BER) pathway.

7.2 OVERVIEW OF THE BASE EXCISION REPAIR PATHWAY

BER is initiated by DNA glycosylases, which recognize damaged and, in some cases, mispaired bases (oxo^{8}dG: adenine [A]) and remove them from DNA by cleaving the N-glycosidic bond between the base and the sugar-phosphate backbone of DNA (Figure 7.7).

Mechanistically, DNA glycosylases can be divided into two classes (Table 7.1) (Dodson et al., 1994): (1) monofunctional DNA glycosylases, which cleave the glycosidic bond using water as a nucleophile and generate apurinic or apyrimidinic (AP) sites, and (2) bifunctional DNA glycosylases/AP lyases, which utilize an amino group of the enzyme as a nucleophile to form a Schiff's base intermediate, which subsequently undergoes enzyme-catalyzed β-elimination cleaving the phosphodiester bond 3' from the abasic site.

Many of the eukaryotic DNA glycosylase mRNAs (*UDG, hOGG1, hNTH1*, and *MYH*) are alternatively spliced to encode nuclear and mitochondrial versions of the protein (Nilsen et al., 1997; Takao et al., 1998; Ohtsubo et al., 2000). Mitochondrial DNA (mtDNA) is predisposed to ROS-induced DNA damage because of the lack of associated structural proteins, e.g., histones; the lack of non-coding segments of DNA, e.g., introns; and its proximity to the ETC. Furthermore, the discovery of mitochondrial BER enzymes has important implications for both actively dividing and terminally differentiated cells since the mitochondrial genome is continuously replicated in all cell types and contains the coding sequences for two ribosomal RNAs (rRNAs), 22 transfer RNAs (tRNAs), and 13 polypeptides (Fernandez-Silva et al., 2003).

All monofunctional glycosylases display activity against damaged cytosine (C) residues. In addition, Mpg is a monofunctional DNA glycosylase that acts on alkylated bases, such as those formed by reaction of lipid peroxidation products with DNA (Wyatt et al., 1999). The bifunctional glycosylases are the first step in the mechanism necessary to repair a host of oxidized DNA base products generated by ROS. hOgg1 and MutY counteract the mutagenic effect of oxo^{8}dG by excising oxo^{8}dG paired with C or A paired to oxo^{8}dG, respectively (Slupska et al., 1999; Boiteux and Radicella, 2000; Ohtsubo et al., 2000). The relevance of maintaining a low basal oxidation state in DNA is underscored by the presence of MTH1, which hydrolyzes oxo^{8}dG triphosphate, thereby preventing the misincorporation of oxidized nucleotide precursors (Sekiguchi, 1996). Together, these three proteins combat the potential for deleterious outcomes from oxo^{8}dG formation. hMTH1 prevents the incorporation of oxo^{8}dG in DNA during DNA replication or repair synthesis by hydrolyzing the oxidized triphosphates. hOgg1 removes oxo^{8}dG from oxo^{8}dG:C base pairs in double-stranded DNA, and MutY removes A from oxo^{8}dG:A pairs, a mispair that may arise solely at sites where oxo^{8}dG was present in the template (Grollman and Moriya, 1993). Additionally, MutY has glycosylase activity towards 8-hydroxy-2'-deoxyadenine (oxo^{2}dA), another mutagenic product of oxidative damage (Ohtsubo et al., 2000). In humans, oxidatively damaged pyrimidine bases such as thymine glycol (Tg) are excised by the hNth1 glycosylase (Dizdaroglu et al., 1999).

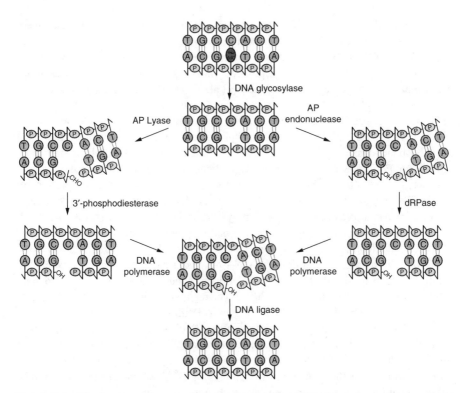

Figure 7.7   The base excision repair pathway is initiated DNA glycosylases that recognize and excise chemically modified bases (oxo⁸dG). DNA glycosylases that possess AP lyase activity cleave the DNA backbone and create 5'-phosphate and 3'-fragmented deoxyribose. The 3'-fragmented deoxyribose is removed by the 3'-diesterase activity of AP endonuclease, generating a one-nucleotide gap with the 5'-phosphate and 3'-OH necessary for DNA synthesis. When DNA glycosylases lack the AP lyase activity, the AP site is incised by AP endonuclease at the 5' side, forming 3'-OH and 5'-abasic deoxyribose phosphate (dRP). The dRP is excised by the dRPase/AP lyase activity of DNA polymerase β, which forms a one-nucleotide gap with 5'-phosphate and 3'-OH. Polymerase β then attaches a single nucleotide to the newly generated 3'-OH and displaces the baseless sugar-phosphate, which it subsequently removed by its inherent AP lyase activity. DNA ligase III seals the nick to restore the original DNA sequence.

Table 7.1   Substrate Specificity of Human DNA Glycosylates

Gene	Protein	Common Substrate	AP Lyase
hOGG1	hOgg1*	oxo⁸dG:C, fapy	Yes
hNTH1	hNth1*	Tg, fapy, oxidized pyrimidines	Yes
MYH	MutY*	A:oxo⁸dG	No
AAG	Mpg	3-MeA, 7-MeG, εA, Hx	No
UNG	Udg*	U	No

Note: Complete names: *hOGG1*, human 8-oxoguanosine DNA glycosylase; *hNTH1*, human endonuclease III homolog; *MYH*, human MutY homolog; *AAG*, human 3-methyladenine DNA glycosylase; *UNG*, uracil DNA glycosylase.

* Nuclear and mitochondrial isoforms have been reported.

Although the action of DNA glycosylases results in the removal of a damaged DNA bases, the resultant AP sites are both cytotoxic and mutagenic. As detailed in Figure 7.7, abasic sites are substrates for AP endonuclease (APE). Following the excision activity of APE, BER can proceed via two pathways, designated as short patch repair or long patch repair (Dianov et al., 1998; Dianov et al., 1992; Dianov et al., 1994; Singhal et al., 1995; Frosina et al., 1996; Krokan et al., 2000).

In the short patch repair pathway, DNA polymerase β adds a single nucleotide and also cleaves the 5'-deoxyribose phosphate residue using an intrinsic deoxyri-bosephosphate lyase (dRP lyase) activity (Bennett et al., 1997b; Matsumoto and Kim, 1995; Podlutsky et al., 2001), and DNA ligase seals the nick to complete repair. However, an AP site can become oxidized or reduced prior to the completion of repair (Biade et al., 1998; Fortini et al., 1998; Podlutsky et al., 2001), thereby ren-dering it resistant to cleavage by DNA polymerase β. This necessitates the removal of the damage through the long patch repair pathway, which begins with the synthesis of a single nucleotide by DNA polymerase β and its subsequent dissociation from the DNA. Polymerase β is replaced by DNA polymerase δ/ε for strand displacement synthesis of up to an additional 10 nucleotides, generating a single-stranded flap (Stucki et al., 1998; Podlutsky et al., 2001). The resultant single-stranded flap enables polymerase β to rebind and complete synthesis (Podlutsky et al., 2001). This flap is removed by flap structure-specific endonuclease-1 (FEN1), thus generating a recog-nizable lesion for DNA ligase to complete repair (Tomkinson et al., 1990). An important characteristic of long patch repair is that it employs many of the same proteins utilized for the replication of the lagging DNA strand during DNA repli-cation, including DNA polymerase δ/ε, replication protein A (RPA), proliferating cellular nuclear antigen (PCNA), FEN1, and DNA ligase I (DeMott et al., 1996; Bambara et al., 1997; Tomkinson and Levin, 1997; Kim et al., 1998; Lippert et al., 1998; Timson et al., 2000). However, alternative interactions have been reported for the completion of long patch repair via poly(ADP-ribose) polymerase-1 (PARP-1)/FEN1/polymerase β and also APE/FEN1/polymerases β/δ/ε (Prasad et al., 2001; Ranalli et al., 2002). It is hypothesized that the former and latter interactions would predominate in post-mitotic cells due to the diminished levels of RPA and PCNA that accompany differentiation. DNA polymerase β is the predominant type in neurons; however, polymerases δ/ε have been shown to be present in appreciable amounts in rat cortical neurons, presumably due to a functional role in long patch repair (Raji et al., 2002). A caveat to be noted with this hypothesis is the uncertainty of the presence of FEN1 or FEN1-like proteins in neurons. Although the expression and function of FEN1 has been well characterized in actively dividing cells, its presence and function in long patch repair in terminally differentiated cells is still unclear. FEN1 expression has been shown to decrease considerably in promyelocytic leukemia cells (HL-60 cells) terminally differentiated into granulocytes or macro-phages (Kim et al., 2000). However, the expression of FEN1 in primary neuronal cultures or a neuron-like cell line, e.g., SHSY5Y neuroblastoma cells or PC12 cells, remains to be determined. Moreover, the repair of oxidative damage in different regions of genome has been demonstrated to be dependent not only on the enzymes involved but also on the types of bases opposite, proximal to, and downstream of the active site of repair.

Previous studies have shown that the N-glycosylase and AP lyase activity of hNth1 and hOgg1 is not concurrent (Zharkov et al., 2000; Marenstein et al., 2001), and several additional proteins can stimulate their activity, including APE (Hill et al., 2001; Vidal et al., 2001; Marenstein et al., 2003). The N-glycosylase activity of hNth1 can be seven times higher than its AP lyase activity when the DNA substrate contained Tg paired to A, whereas the two activities are comparable when Tg is paired to G (Marenstein et al., 2001; Ocampo et al., 2002). The base pairing to Tg residue in DNA also modulates the stimulatory action of APE on hNth1 (Marenstein et al., 2003). In contrast, the β-elimination activity of hNth1 is strongly dependent upon the nature of the base pairing with the AP site, being 100 times greater with AP:G as compared to AP:A (Eide et al., 2001). These findings are indicative of the involvement of different substrate-dependent interacting and modifying proteins *in vivo*, since it has been suggested that the intrinsic properties of the glycosylase involved in the initial recognition and removal of the damaged base may be responsible for BER pathway selection (Fortini et al., 1999). Finally, the position of oxo^{8}dG in relation to AP sites has been shown to impair not only strand displacement synthesis by DNA polymerases β/δ but also the cleavage capacity of FEN1, when oxo^{8}dG is located one nucleotide 3'-downstream of the AP site (Budworth et al., 2002).

7.3 NEURONAL VULNERABILITY ASSOCIATED WITH DEFECTIVE DNA REPAIR

Several knockout mice lacking BER enzymes have been generated using gene-targeting methods. A striking observation from these studies is that mice deficient in DNA glycosylases (Mpg, hOgg1, Udg, Nth1) show no overt phenotypic abnormalities and have no predisposition to cancer, while mice deficient in genes involved in BER downstream of the DNA glycosylases (APE, Polymerase β, and XRCC1) all show embryonic lethality (Engelward et al., 1997; Hang et al., 1997; Wilson and Thompson, 1997; Klungland et al., 1999; Minowa et al., 2000; Nilsen et al., 2000). Numerous protein-protein interactions have been identified between the BER and NER pathways, indicating the highly coordinated nature of these processes (Bennett et al., 1997a; Mol et al., 2000; Waters et al., 1999). The implications of the overlapping roles of BER and NER have shed some light on the similarities and neuro-pathological features of individuals afflicted with genetic disorders in which the defect is associated with a mutation in enzymes that participate in DNA repair.

Xeroderma pigmentosum (XP) and Cockayne's syndrome (CS) are two representative genetic disorders that have revealed the importance of functional DNA repair pathways at maintaining neuronal stability (Cleaver and Crowley, 2002; Rapin et al., 2000). XP is a rare recessively inherited disorder caused by a mutation in one of several genes encoding proteins required for normal repair of DNA damage induced by ultraviolet radiation (UV) or UV-mimetic agents. Currently, there are seven identified complementation groups for XP (XP-A to -G) and a variant form, which is characterized by a defective polymerase (Moriwaki and Kraemer, 2001). XP patients experience a greater than 1000-fold excess frequency of sunlight-related skin cancers. In normal people there is an increased risk of skin cancer with aging, and this correlates with age-related declines in DNA repair. Secondary to skin cancers, approximately 20–30% of XP patients suffer from neurological impairment, resulting from a decreased capacity to repair oxidative DNA lesions, namely, oxo^{8}dG and Tg (Reardon et al., 1997). Moreover, XPG (XP, Group G protein) has been shown to facilitate the removal of these lesions by enhancing the activity of Ogg1 and hNth1, respectively (Klungland et al., 1999; Le Page et al., 2000).

CS has two complementation groups, CSA and CSB. Clinically, CS patients exhibit premature aging due to defective repair of actively transcribed genes that lead to sun sensitivity, dwarfism, retinal degeneration, progressive sensorineural hearing loss, and accelerated neurological degeneration. Both CSA and CSB have been ascribed a function in the TCR subpathway of NER that results in the removal of bulky lesions from the transcribed strand of RNA polymerase II transcribed genes, thus permitting the rapid resumption of blocked transcription (Svejstrup, 2002). Interestingly, oxo^{8}dG has been shown to block transcription mediated by RNA polymerase II in shuttle vectors containing a single oxo^{8}dG:C base pair (Le Page et al., 2000), and CS cells accumulate oxo^{8}dG due to a reduction in *hOGG1* expression (Dianov et al., 1999). Moreover, the integrity of the CSB protein has been shown to play a pivotal role in the steps preceding the removal of oxo^{8}dG, and it appears that a functional CSB gene improves the processing of this lesion during TCR and is essential for efficient removal of this lesion during GGR (Sunesen et al., 2002; Tuo et al., 2002).

Such clinical correlations underscore the importance of DNA repair mechanisms at maintaining genomic stability and preventing neuronal degeneration. However, the aforementioned diseases are characterized by heritable defects, whereas the majority of cases of age-associated neurological deficits such as Parkinson's disease (PD), Alzheimer's disease (AD), and amyotrophic lateral sclerosis (ALS) are sporadic in nature, and despite a predominance of oxidative DNA lesions found in postmortem tissues of disease sufferers (Sanchez-Ramos et al., 1994; Bogdanov et al., 2000; Lovell et al., 2000), no genetic deficits in DNA repair have been identified, suggesting that these diseases may result from environmental toxicants or behavioral factors that predispose an individual to oxidative stressors (Fleming et al., 1994; Khabazian et al., 2002; Mattson et al., 2002).

7.4 CHEMICAL EXPOSURES, OXIDATIVE STRESS, AND NEURODEGENERATION

The environmental toxin/oxidative stress theory of neurodegeneration has been most strongly supported by the findings that several neurotoxicants, e.g., 1-methyl-4-phenyl-1,2,3,6-tetrahydropyridine (MPTP), dieldrin, paraquat, and rotenone are capable of eliciting neurodegeneration by disrupting various complexes in the ETC and increasing ROS production (Betarbet et al., 2000; Corrigan et al., 2000; Fleming et al., 1994). Additionally, mitochondrial abnormalities have been identified in a large proportion of individuals afflicted with neurodegenerative diseases (Beal, 1998), including complex I deficiency in the substantia nigra of patients with sporadic PD (Mizuno et al., 1989; Parker et al., 1989; Schapira et al., 1989); impaired cytochrome c oxidase activity in the CNS and peripheral tissues of AD patients (Kish et al., 1992; Mutisya et al., 1994; Parker et al., 1994); and mutations in the mtDNA-encoded subunit of cytochrome c oxidase in patients with sporadic ALS (Comi et al., 1998). However, whether the aforementioned chemicals or mitochondrial abnormalities are primary and related to the causation of the disease process or surrogate markers of some other deficiency is still unresolved. It is possible that chronic exposure to specific neurotoxicants or behavioral factors, e.g., excessive caloric intake, may preclude an individual to a higher degree of oxidative stress, which results in acquired mutations in the mtDNA. Mutations within the mitochondrial genome can accumulate within individual cells, and eventually compromise mitochondrial function (Hofhaus et al., 2003).

Chronic low-level exposures to neurotoxicants and caloric restriction have been shown to have profound effects on neurodegeneration. For instance, chronic rotenone exposure results in a uniform inhibition of complex I throughout the brain; however, despite this nonselective inhibition, rotenone causes a selective degeneration in neurons predisposed to high levels of oxidative stress, dopaminergic neurons of the nigrostriatal pathway (Betarbet et al., 2000). On the other hand, caloric restriction has been shown to attenuate age-associated changes in mitochondrial proteins that prevent apoptosis (Shelke and Leeuwenburgh, 2003). Furthermore, caloric restriction has been shown to significantly decrease the MPTP-induced loss of dopaminergic neurons and motor function in rats and mice (Duan and Mattson, 1999).

Human populations typically display a range of inherent sensitivities to radiation and chemical-induced cellular changes. Given a common chemical insult, some individuals develop associated health problems, whereas others remain clinically free from all related effects of exposure (Lockridge and Masson, 2000). Even within specific populations of exposed individuals such as pesticide applicators, the age of onset and the extent and severity of neurological problems often vary among individuals (Cantor and Silberman, 1999). Such variability in host response may be due, in part, to inherent differences between individuals in the levels of oxidative stress and their ability to monitor and repair damaged sites induced in their genetic material by exogenous and endogenous agents.

This type of variability in response is commonly encountered in the laboratory setting as well, as the choice of animal or strain of animal can have a dramatic impact on the observed outcome for chemical exposures (Kacew et al., 1998; Kacew et al., 1995). For example, when rats are administered with mg/kg doses of MPTP comparable to those used in mice, rats do not exhibit any significant dopaminergic degeneration (Giovanni et al., 1994). The optimal neurotoxic effects of MPTP are seen in the C57BL/6J strain of mouse (Przedborski et al., 2001). When Balb/c and C57BL/6J mice are treated with MPTP, neostriatal dopamine depletion is greatest in the C57BL/6J strain (-85%) versus the Balb/cJ strain (-58%). Furthermore, a loss of tyrosine hydroxylase immunoreactivity is only observed in the substantia nigra of C57BL/6J animals (–25%) (Sedelis et al., 2000). The apparent susceptibility of this strain was initially attributed to a greater density of monoamine oxidase type-B in the basal ganglia and substantia nigra, rather than a differential distribution of the enzyme, compared to other strains (Zimmer and Geneser, 1987). This may account for an increased capacity for bioactivating MPTP to its neurotoxic metabolite 1-methyl-4-phenylpyridinium (MPP+); however, it fails to explain why C57BL/6J mice are more susceptible to MPTP toxicity, considering the tissue concentration of MPP+ does not appear to be the determining factor for vulnerability of dopaminergic and noradrenergic neurons to MPTP. High concentrations of MPP+ have been detected in the striatum of brown Norway rats without degenerative effects on dopaminergic neurons (Zuddas et al., 1994). Additional studies have shown that there is not a significant difference in the capacity of neostriatal synaptosomes to sequester MPP+ between Ace Swiss-Webster, CD-1, or C57BL/6J mice. (Giovanni et al., 1991).

The differences observed between animal species or strains of mice to MPTP toxicity may be explained by regional or cellular differences in antioxidant or DNA repair enzymes, as recent evidence has shown that the levels of oxo^8dG in different brain regions of C57BL/6J mice inversely correlated with the activity of Ogg1 (Cardozo-Pelaez et al., 2000). Although this relationship was observed in tissue homogenates containing both actively dividing and differentiated cells, it has since been determined that an inverse relationship between the activity of Ogg1 and the levels of oxo^8dG exists in terminally differentiated PC 12 cells challenged with the organochlorine pesticide, dieldrin; however, this relationship was not observed in actively dividing cells (Stedeford et al., 2001). This difference in response may be attributed to the higher basal levels of Ogg1 activity in actively dividing cells, since there is a decrease in both Ogg1 protein expression and activity, and a corresponding increase in oxo^8dG levels and Mn-SOD activity (F. Cardozo-Pelaez, unpublished findings), when PC 12 cells are differentiated with nerve growth factor (Figure 7.8). The reader is referred to Cardozo-Pelaez et al. (2000) for a detailed explanation of the experimental protocols utilized for Figures 7.8–7.13.

Subsequent studies with C57BL/6J mice have indicated that Ogg1 activity is inducible in specific brain regions with low basal levels of activity (cerebellum, cortex, and pons/medulla) and remains unchanged in those regions with high basal levels of activity (hippocampus, caudate/putamen, and midbrain) (Cardozo-Pelaez et al., 2002). Comparison studies with three different mouse strains (ICR, Balb/cJ, and C57BL/6J) of a region with low Ogg1 activity (cerebellum) versus a region with high Ogg1 activity (caudate/putamen) have revealed that although no region or strain-specific differences exist in oxo^8dG levels between the ICR and Balb/cJ strains, the C57BL/6J strain has significantly higher levels of oxo^8dG in the cerebellum (1.98- and 2.2-fold versus Balb/c and ICR, respectively) and caudate putamen (1.82- and 2.7-fold versus Balb/c and ICR, respectively) (Figure 7.9).

Regional analysis of Ogg1 activity in the cerebellum of male ICR, Balb/c, and C57BL/6J mice has not revealed any significant differences. However, ICR mice have a 66% lower level of Ogg1 activity in the caudate putamen (Figure 7.10), yet no differences have been observed in the protein levels of Ogg1 by mouse strain or by brain region (Figure 7.11).

As mentioned previously, superoxide dismutases can generate H_2O_2, which can transverse cellular membranes and cause increases in both mitochondrial and nuclear oxo^8dG levels. The activities of the cytosolic Cu/Zn-SOD and the mitochondrial Mn-SOD have been shown to differ significantly by mouse strain and by brain region.

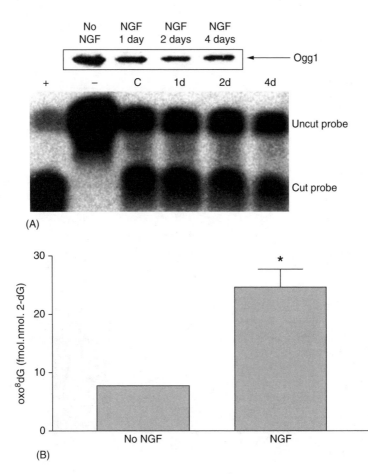

(A)

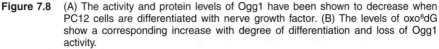

(B)

Figure 7.8 (A) The activity and protein levels of Ogg1 have been shown to decrease when PC12 cells are differentiated with nerve growth factor. (B) The levels of oxo^{8}dG show a corresponding increase with degree of differentiation and loss of Ogg1 activity.

Cu/Zn-SOD is significantly elevated by greater than 2.45-fold in both the cerebellum and caudate putamen of C57BL/6J mice versus Balb/cJ and ICR mice (Figure 7.12). In contrast, no significant differences in Mn-SOD activity exists between strains in the cerebellum; however, the C57BL/6J strain has a greater than 6.65-fold increase in Mn-SOD activity in the caudate putamen compared to Balb/c and ICR mice (Figure 7.13). A difference that is not reflective of a decreased level of Ogg1 activity or protein expression in C57BL/6J mice versus ICR or Balb/cJ, despite the enhanced activities of Mn- and Cu/Zn-SOD in the cerebellum and caudate/putamen of C57BL/6J mice.

It is suggested that the enhanced susceptibility of the C57BL/J strain to neurotoxicants (e.g., MPTP) may be due to saturation of BER pathways. In mitochondria, $O_2^{\cdot-}$ is the primary product of unreduced oxygen, which results in the production of H_2O_2 either enzymatically by Mn-SOD or by spontaneous disproportionation. Inhi-

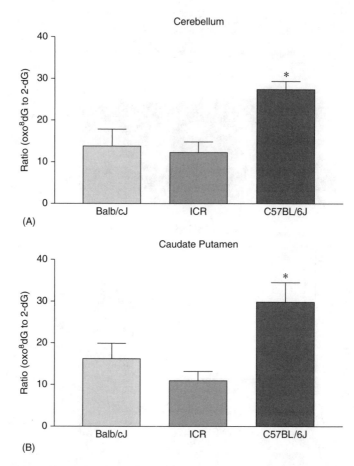

Figure 7.9 Oxo8dG levels in the cerebellum and caudate putamen of Balb/cJ, ICR, and C57BL/6J mice. The levels of oxo8dG in both the (A) cerebellum and (B) caudate putamen were significantly higher in the C57BL/6J strain than Balb/cJ or ICR. Data are expressed as the mean ± S.E.M., n = 4. An * indicates a statistically significant difference, p <0.05. (From Mosquera, D.I. et al. (2003). *Journal of Gene Expression* 11(1), 47–53. With permission.)

bition of complex I by MPP⁺ has been shown to result in an increased production of superoxide anion. Therefore, the enhanced susceptibility of C57BL/6J mice to MPTP may be the result of the high basal activity of Mn-SOD, as it would serve to increase the production of H_2O_2, and ultimately, levels of oxidative DNA damage.

The increased activity of Cu/Zn-SOD in C57BL/6J mice further supports this notion, as enhanced Cu/Zn-SOD activity has been linked with Down's syndrome, a disease characterized by premature aging and neurological impairment (Brooksbank and Balazs, 1984). In neural tissues, Cu/Zn-SOD and glutathione peroxidase are the predominant oxygen scavenging enzymes (Jain et al., 1991b; Marklund et al., 1983b). A linear increase in Cu/Zn-SOD activity has been reported with aging in mouse brains; however, there was no detectable increase in glutathione peroxidase

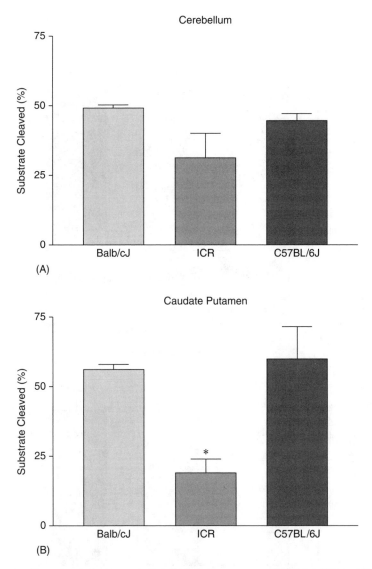

(A)

(B)

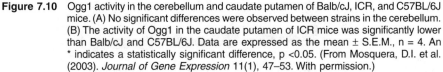

Figure 7.10 Ogg1 activity in the cerebellum and caudate putamen of Balb/cJ, ICR, and C57BL/6J mice. (A) No significant differences were observed between strains in the cerebellum. (B) The activity of Ogg1 in the caudate putamen of ICR mice was significantly lower than Balb/cJ and C57BL/6J. Data are expressed as the mean ± S.E.M., n = 4. An * indicates a statistically significant difference, p <0.05. (From Mosquera, D.I. et al. (2003). *Journal of Gene Expression* 11(1), 47–53. With permission.)

with aging, as is the case in neural tissues of Down's syndrome patients (de Haan et al., 1992).

Since the striatum is predisposed to oxidative stress from dopamine metabolism and auto-oxidation, the enhanced activity of SODs in this region of the C57BL/J strain may preclude its susceptibility to neurotoxicants. It is interesting to note,

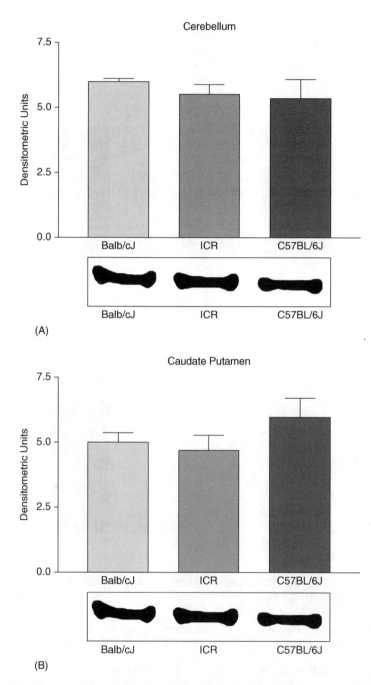

Figure 7.11 Western blot analysis of Ogg1 in the cerebellum and caudate putamen of Balb/cJ, ICR, and C57BL/6J mice. No significant differences are present between (A) cerebellum or the (B) caudate putamen. Data are expressed as the mean ± S.E.M., n = 4. (From Mosquera, D.I. et al. (2003). *Journal of Gene Expression* 11(1), 47–53. With permission.)

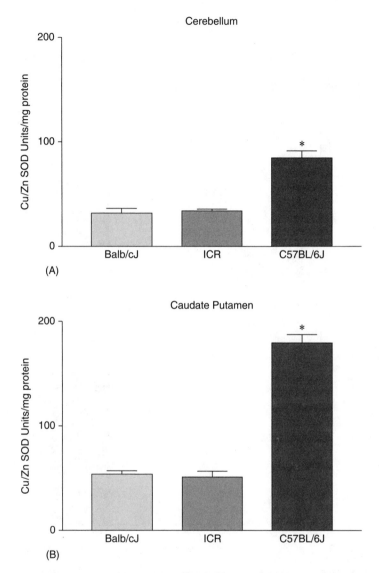

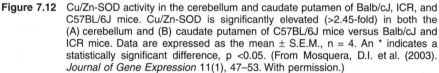

Figure 7.12 Cu/Zn-SOD activity in the cerebellum and caudate putamen of Balb/cJ, ICR, and C57BL/6J mice. Cu/Zn-SOD is significantly elevated (>2.45-fold) in both the (A) cerebellum and (B) caudate putamen of C57BL/6J mice versus Balb/cJ and ICR mice. Data are expressed as the mean ± S.E.M., n = 4. An * indicates a statistically significant difference, p <0.05. (From Mosquera, D.I. et al. (2003). *Journal of Gene Expression* 11(1), 47–53. With permission.)

however, that the ICR strain of mice have previously been shown to have a higher brain dopamine concentration versus the Balb/c and C57BL/J strain (Messiha et al., 1990), suggesting that the enhanced activities of SODs in the C57BL/J mice may not be reflective of a compensatory response to increased levels of ROS generated from the metabolism and auto-oxidation of dopamine.

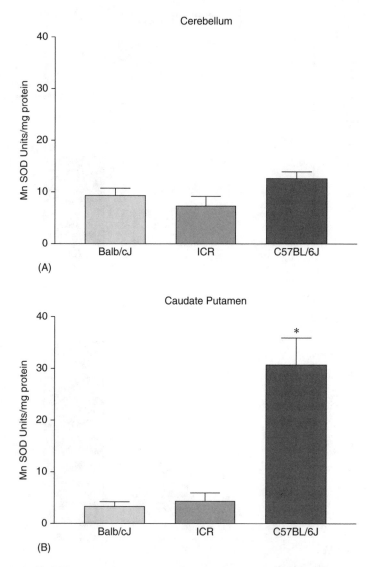

Figure 7.13 MnSOD activity in the cerebellum and caudate putamen of Balb/cJ, ICR, and C57BL/6J mice. No significant differences in MnSOD activity exists between strains in the (A) cerebellum; however, the C57BL/6J strain possesses a significantly higher (>6.65-fold) activity in the caudate putamen compared to Balb/c and ICR mice. Data are expressed as the mean ± S.E.M., n = 4. An * indicates a statistically significant difference, p <0.05. (From Mosquera, D.I. et al. (2003). *Journal of Gene Expression* 11(1), 47–53. With permission.)

7.5 FUTURE PERSPECTIVES

As presented herein, a complete understanding of DNA damage and DNA repair in the CNS is in the developmental stages. Several of the enzymes involved in BER for short patch and long patch repair of actively dividing cells have yet to be identified in terminally differentiated cells. Until a more complete understanding of the enzymes involved in neuronal cell types is understood, the effects of the bases surrounding the active site of repair will remain elusive. Furthermore, several key processes involved in the down-regulation of gene expression, e.g., methylation, may alter the processing of damaged bases by the BER pathway and may be more relevant with regards to BER processes than previously expected. The enzyme DNA methyltransferase 1 (DNMT1) is responsible for maintaining DNA C methylation, and is essential for survival (Fan et al., 2001). Methylation of C residues (5-methylcytosine) is a fundamental process for silencing gene transcription in regions of simple dinucleotide sites known as CpG islands (Ng and Bird, 1999). The majority of unmethylated CpG sites are located near the promoter region and first exon region of protein coding genes (Lander et al., 2001; Venter et al., 2001). Therefore, the generation of Tg from oxidation of methylated bases such as 5-methylcytosine may result in the incorporation of a C that will not be remethylated and may eventually lead to regions of hypomethylated DNA, which can cause neurodegeneration (Fan et al., 2001). In addition, the activity of DNMT1 has been shown to decrease abruptly when PC12 cells are differentiated with nerve growth factor (Deng and Szyf, 1999), so the interplay between DNA damage, repair, and methylation capacity in neurons may have profound significance in neurodegeneration. Finally, as discussed in the latter section, a better understanding of repair processes in test animals is of fundamental importance for advancing the understanding of DNA damage and DNA repair and their relation to the mechanisms that regulate these processes in humans.

REFERENCES

Ames, B. N. (1983), Dietary carcinogens and anticarcinogens. Oxygen radicals and degenerative diseases, *Science* 221, 1256–1264.

Ames, B. N. (1989), Endogenous DNA damage as related to cancer and aging, *Mutat. Res.* 214, 41–46.

Ames, B. N., Shigenaga, M. K., and Hagen, T. M. (1993), Oxidants, antioxidants, and the degenerative diseases of aging, *Proc. Natl. Acad. Sci. USA* 90, 7915–7922.

Bambara, R. A., Murante, R. S., and Henricksen, L. A. (1997), Enzymes and reactions at the eukaryotic DNA replication fork, *J. Biol. Chem.* 272, 4647–4650.

Beal, M. F. (1998), Mitochondrial dysfunction in neurodegenerative diseases, *Biochim. Biophys. Acta* 1366, 211–223.

Bennett, R. A. et al. (1997a), Interaction of human apurinic endonuclease and DNA polymerase beta in the base excision repair pathway, *Proc. Natl. Acad. Sci. USA* 94, 7166–7169.

Bennett, R. A. et al. (1997b), Interaction of human apurinic endonuclease and DNA polymerase beta in the base excision repair pathway, *Proc. Natl. Acad. Sci. USA* 94, 7166–7169.

Betarbet, R. et al. (2000), Chronic systemic pesticide exposure reproduces features of Parkinson's disease, *Nat. Neurosci.* 3, 1301–1306.

Biade, S. et al. (1998), Impairment of proliferating cell nuclear antigen-dependent apurinic/apyrimidinic site repair on linear DNA, *J. Biol. Chem.* 273, 898–902.

Bogdanov, M. et al. (2000), Increased oxidative damage to DNA in ALS patients, *Free Radic. Biol. Med.* 29, 652–658.

Boiteux, S., and Radicella, J. P. (2000), The human OGG1 gene: structure, functions, and its implication in the process of carcinogenesis, *Arch. Biochem. Biophys.* 377, 1–8.

Breimer, L. H. (1991), Repair of DNA damage induced by reactive oxygen species, *Free Radic. Res. Commun.* 14, 159–171.

Brooksbank, B. W., and Balazs, R. (1984), Superoxide dismutase, glutathione peroxidase and lipoperoxidation in Down's syndrome fetal brain, *Brain Res.* 318, 37–44.

Budworth, H. et al. (2002), Repair of clustered DNA lesions. Sequence-specific inhibition of long-patch base excision repair be 8-oxoguanine, *J. Biol. Chem.* 277, 21300–21305.

Cantor, K. P., and Silberman, W. (1999), Mortality among aerial pesticide applicators and flight instructors: follow-up from 1965-1988, *Am. J. Ind. Med.* 36, 239–247.

Cardozo-Pelaez, F. et al. (2000), DNA damage, repair, and antioxidant systems in brain regions: a correlative study, *Free Radic. Biol. Med.* 28, 779–785.

Cardozo-Pelaez, F. et al. (2002), Effects of diethylmaleate on DNA damage and repair in the mouse brain, *Free Radic. Biol. Med.* 33, 292–298.

Cleaver, J. E., and Crowley, E. (2002), UV damage, DNA repair and skin carcinogenesis, *Front. Biosci.* 7, d1024–1043.

Comi, G. P. et al. (1998), Cytochrome c oxidase subunit I microdeletion in a patient with motor neuron disease, *Ann. Neurol.* 43, 110–116.

Comporti, M. (1985), Lipid peroxidation and cellular damage in toxic liver injury, *Lab Invest.* 53, 599–623.

Corrigan, F. M. et al. (2000), Organochlorine insecticides in substantia nigra in Parkinson's disease, *J. Toxicol. Environ. Health* A 59, 229–234.

de Haan, J. B., Newman, J. D., and Kola, I. (1992), Cu/Zn superoxide dismutase mRNA and enzyme activity, and susceptibility to lipid peroxidation, increases with aging in murine brains, *Brain Res. Mol. Brain Res.* 13, 179–187.

de Laat, W. L., Jaspers, N. G., and Hoeijmakers, J. H. (1999), Molecular mechanism of nucleotide excision repair, *Genes Dev.* 13, 768–785.

DeMott, M. S. et al. (1996), Human RAD2 homolog 1 5'- to 3'-exo/endonuclease can efficiently excise a displaced DNA fragment containing a 5'-terminal abasic lesion by endonuclease activity, *J. Biol. Chem.* 271, 30068–30076.

Deng, J., and Szyf, M. (1999), Downregulation of DNA (cytosine-5-methyltransferase) is a late event in NGF-induced PC12 cell differentiation, *Brain Res. Mol. Brain Res.* 71, 23–31.

Dianov, G., Price, A., and Lindahl, T. (1992), Generation of single-nucleotide repair patches following excision of uracil residues from DNA, *Mol. Cell Biol.* 12, 1605–1612.

Dianov, G. et al. (1994), Release of 5'-terminal deoxyribose-phosphate residues from incised abasic sites in DNA by the Escherichia coli RecJ protein, *Nucleic Acids Res.* 22, 993–998.

Dianov, G. et al. (1998), Repair pathways for processing of 8-oxoguanine in DNA by mammalian cell extracts, *J. Biol. Chem.* 273, 33811–33816.

Dianov, G. et al. (1999), Repair of 8-oxoguanine in DNA is deficient in Cockayne syndrome group B cells, *Nucleic Acids Res.* 27, 1365–1368.

Dizdaroglu, M. et al. (1999), Excision of products of oxidative DNA base damage by human NTH1 protein, *Biochemistry* 38, 243-246.

Dodson, M. L., Michaels, M. L., and Lloyd, R. S. (1994), Unified catalytic mechanism for DNA glycosylases, *J. Biol. Chem.* 269, 32709–32712.

Duan, W., and Mattson, M. P. (1999), Dietary restriction and 2-deoxyglucose administration improve behavioral outcome and reduce degeneration of dopaminergic neurons in models of Parkinson's disease, *J. Neurosci. Res.* 57, 195–206.

Eide, L. et al. (2001), Human endonuclease III acts preferentially on DNA damage opposite guanine residues in DNA, *Biochemistry* 40, 6653–6659.

Engelward, B. P. et al. (1997), Base excision repair deficient mice lacking the Aag alkyladenine DNA glycosylase, *Proc. Natl. Acad. Sci. USA* 94, 13087–13092.

Fan, G. et al. (2001), DNA hypomethylation perturbs the function and survival of CNS neurons in postnatal animals, *J. Neurosci.* 21, 788–797.

Fernandez-Silva, P., Enriquez, J. A., and Montoya, J. (2003), Replication and transcription of mammalian mitochondrial DNA, *Exp. Physiol.* 88, 41–56.

Fleming, L. et al. (1994), Parkinson's disease and brain levels of organochlorine pesticides, *Ann. Neurol.* 36, 100–103.

Fortini, P. et al. (1999), The type of DNA glycosylase determines the base excision repair pathway in mammalian cells, *J. Biol. Chem.* 274, 15230–15236.

Fortini, P. et al. (1998), Different DNA polymerases are involved in the short- and long-patch base excision repair in mammalian cells, *Biochemistry* 37, 3575–3580.

Fridovich, I. (1978), The biology of oxygen radicals, *Science* 201, 875–880.

Frosina, G. et al. (1996), Two pathways for base excision repair in mammalian cells, *J. Biol. Chem.* 271, 9573–9578.

Friedberg, E. C. (1995), Out of the shadows and into the light: the emergence of DNA repair, *Trends Biochem. Sci.* 20, 381.

Fucci, L. et al. (1983), Inactivation of key metabolic enzymes by mixed-function oxidation reactions: possible implication in protein turnover and ageing, *Proc. Natl. Acad. Sci. USA* 80, 1521–1525.

Giovanni, A. et al. (1991), Correlation between the neostriatal content of the 1-methyl-4-phenylpyridinium species and dopaminergic neurotoxicity following 1-methyl-4-phenyl-1,2,3,6-tetrahydropyridine administration to several strains of mice, *J. Pharmacol. Exp. Ther.* 257, 691–697.

Giovanni, A. et al. (1994), Studies on species sensitivity to the dopaminergic neurotoxin 1-methyl-4-phenyl-1,2,3,6-tetrahydropyridine. Part 1: Systemic administration, *J. Pharmacol. Exp. Ther.* 270, 1000–1007.

Grollman, A. P., and Moriya, M. (1993), Mutagenesis by 8-oxoguanine: an enemy within, *Trends Genet.* 9, 246–249.

Halliwell, B. (1994), Free radicals and antioxidants: a personal view. *Nutr. Rev.* 52, 253–265.

Hang, B. et al. (1997), Targeted deletion of alkylpurine-DNA-N-glycosylase in mice eliminates repair of 1,N6-ethenoadenine and hypoxanthine but not of 3,N4-ethenocytosine or 8-oxoguanine, *Proc. Natl. Acad. Sci. USA* 94, 12869–12874.

Harman, D. (1981), The aging process, *Proc. Natl. Acad. Sci. USA* 78, 7124–7128.

Harman, D. (1993), Free radical involvement in aging. Pathophysiology and therapeutic implications, *Drugs Aging* 3, 60–80.

Hill, J. W. et al. (2001), Stimulation of human 8-oxoguanine-DNA glycosylase by AP-endonuclease: potential coordination of the initial steps in base excision repair, *Nucleic Acids Res.* 29, 430–438.

Hofhaus, G. et al. (2003), Live now—pay by ageing: high performance mitochondrial activity in youth and its age-related side effects, *Exp. Physiol.* 88, 167–174.

Jain, A. et al. (1991a), Glutathione deficiency leads to mitochondrial damage in brain, *Proc. Natl. Acad. Sci. USA* 88, 1913–1917.

Jain, A. et al. (1991b), Glutathione deficiency leads to mitochondrial damage in brain, *Proc. Natl. Acad. Sci. USA* 88, 1913–1917.

Kacew, S., Ruben, Z., and McConnell, R. F. (1995), Strain as a determinant factor in the differential responsiveness of rats to chemicals, *Toxicol. Pathol.* 23, 701–714..

Kacew, S., Dixit, R., and Ruben, Z. (1998), Diet and rat strain as factors in nervous system function and influence of confounders, *Biomed. Environ. Sci.* 11, 203–217.

Khabazian, I. et al. (2002), Isolation of various forms of sterol beta-D-glucoside from the seed of Cycas circinalis: neurotoxicity and implications for ALS-parkinsonism dementia complex, *J. Neurochem.* 82, 516–528.

Kim, I. S. et al. (2000), Gene expression of flap endonuclease-1 during cell proliferation and differentiation, *Biochem. Biophys. Acta* 1496, 333–340.

Kim, K., Biade, S., and Matsumoto, Y. (1998), Involvement of flap endonuclease 1 in base excision DNA repair, *J. Biol. Chem.* 273, 8842–8848.

Kish, S. J. et al. (1992), Brain cytochrome oxidase in Alzheimer's disease, *J. Neurochem.* 59, 776–779.

Klungland, A. et al. (1999), Base excision repair of oxidative DNA damage activated by XPG protein, *Mol. Cell* 3, 33–42.

Koppenol, W. H. (1990), What is in a name? Rules for radicals, *Free Radic. Biol. Med.* 9, 225–227.

Krokan, H. E. et al. (2000), Base excision repair of DNA in mammalian cells, *FEBS Lett.* 476, 73–77.

Lander, E. S. et al. (2001), Initial sequencing and analysis of the human genome, *Nature* 409, 860–921.

Le Page, F. et al. (2000), Transcription-coupled repair of 8-oxoguanine: requirement for XPG, TFIIH, and CSB and implications for Cockayne syndrome, *Cell* 101, 159–171.

Lindahl, T., and Nyberg, B. (1972), Rate of depurination of native deoxyribonucleic acid, *Biochemistry* 11, 3610–3618.

Lippert, M. J., Chen, Q., and Liber, H. L. (1998), Increased transcription decreases the spontaneous mutation rate at the thymidine kinase locus in human cells, *Mutat. Res.* 401, 1–10.

Lockridge, O., and Masson, P. (2000), Pesticides and susceptible populations: people with butyrylcholinesterase genetic variants may be at risk, *NeuroToxicology* 21, 113–126.

Lovell, M. A., Xie, C., and Markesbery, W. R. (2000), Decreased base excision repair and increased helicase activity in Alzheimer's disease brain, *Brain Res.* 855, 116–123.

Marenstein, D. R. et al. (2001), Stimulation of human endonuclease III by Y box-binding protein 1 (DNA-binding protein B). Interaction between a base excision repair enzyme and a transcription factor, *J. Biol. Chem.* 276, 21242–21249.

Marenstein, D. R. et al. (2003), Substrate specificity of human endonuclease III (hNth1): effect of human AP endonuclease 1 (APE1) on hNth1 activity, *J Biol. Chem.*

Marklund, S. L. et al. (1983a), Superoxide dismutase activity in brains from chronic alcoholics, *Drug Alcohol Depend.* 12, 209–215.

Marklund, S. L. et al. (1983b), Superoxide dismutase activity in brains from chronic alcoholics, *Drug Alcohol Depend.* 12, 209–215.

Matsumoto, Y., and Kim, K. (1995), Excision of deoxyribose phosphate residues by DNA polymerase beta during DNA repair, *Science* 269, 699–702.

Mattson, M. P., Chan, S. L., and Duan, W. (2002), Modification of brain aging and neurodegenerative disorders by genes, diet, and behavior, *Physiol. Rev.* 82, 637–672.

Messiha, F. S., Martin, W. J., and Bucher, K. D. (1990), Behavioral and genetic interrelationships between locomotor activity and brain biogenic amines, *Gen. Pharmacol.* 21, 459–464.

Minowa, O. et al. (2000), Mmh/Ogg1 gene inactivation results in accumulation of 8-hydroxyguanine in mice, *Proc. Natl. Acad. Sci. USA* 97, 4156–4161.

Mizuno, Y. et al. (1989), Deficiencies in complex I subunits of the respiratory chain in Parkinson's disease, *Biochem. Biophys. Res. Commun.* 163, 1450–1455.

Mol, C. D. et al. (2000), DNA-bound structures and mutants reveal abasic DNA binding by APE1 and DNA repair coordination [corrected], *Nature* 403, 451–456.

Moosmann, B., and Behl, C. (2002), Antioxidants as treatment for neurodegenerative disorders, *Expert. Opin. Investig. Drugs* 11, 1407–1435.

Moriwaki, S., and Kraemer, K. H. (2001), Xeroderma pigmentosum — bridging a gap between clinic and laboratory, *Photodermatol. Photoimmunol. Photomed.* 17, 47–54.

Mutisya, E. M., Bowling, A. C., and Beal, M. F. (1994), Cortical cytochrome oxidase activity is reduced in Alzheimer's disease, *J. Neurochem.* 63, 2179–2184.

Ng, H. H., and Bird, A. (1999), DNA methylation and chromatin modification, *Curr. Opin. Genet. Dev.* 9, 158–163.

Nilsen, H. et al. (1997), Nuclear and mitochondrial uracil-DNA glycosylases are generated by alternative splicing and transcription from different positions in the UNG gene, *Nucleic. Acids Res.* 25, 750–755.

Nilsen, H. et al. (2000), Uracil-DNA glycosylase (UNG)-deficient mice reveal a primary role of the enzyme during DNA replication, *Mol. Cell* 5, 1059–1065.

Nouspikel, T., and Hanawalt, P. C. (2000), Terminally differentiated human neurons repair transcribed genes but display attenuated global DNA repair and modulation of repair gene expression, *Mol. Cell Biol.* 20, 1562–1570.

Nouspikel, T., and Hanawalt, P. C. (2002), DNA repair in terminally differentiated cells, *DNA Repair (Amst.)* 1, 59–75.

Ocampo, M. T. et al. (2002), Targeted deletion of mNth1 reveals a novel DNA repair enzyme activity, *Mol. Cell Biol.* 22, 6111–6121.

Ohtsubo, T. et al. (2000), Identification of human MutY homolog (hMYH) as a repair enzyme for 2-hydroxyadenine in DNA and detection of multiple forms of hMYH located in nuclei and mitochondria, *Nucleic Acids Res.* 28, 1355–1364.

Papa, S. (1996), Mitochondrial oxidative phosphorylation changes in the life span. Molecular aspects and physiopathological implications, *Biochem. Biophys. Acta* 1276, 87–105.

Parker, W. D., Jr., Boyson, S. J., and Parks, J. K. (1989), Abnormalities of the electron transport chain in idiopathic Parkinson's disease, *Ann. Neurol.* 26, 719–723.

Parker, W. D., Jr. et al. (1994), Electron transport chain defects in Alzheimer's disease brain, *Neurology* 44, 1090–1096.

Podlutsky, A. J. et al. (2001), Human DNA polymerase beta initiates DNA synthesis during long-patch repair of reduced AP sites in DNA, *EMBO. J.* 20, 1477–1482.

Prasad, R. et al. (2001), DNA polymerase beta -mediated long patch base excision repair. Poly(ADP-ribose)polymerase-1 stimulates strand displacement DNA synthesis, *J. Biol. Chem.* 276, 32411–32414.

Przedborski, S. et al. (2001), The parkinsonian toxin 1-methyl-4-phenyl-1,2,3,6-tetrahydro-pyridine (MPTP): a technical review of its utility and safety, *J. Neurochem.* 76, 1265–1274.

Raji, N. S., Krishna, T. H., and Rao, K. S. (2002), DNA-polymerase alpha, beta, delta and epsilon activities in isolated neuronal and astroglial cell fractions from developing and aging rat cerebral cortex, *Int. J. Dev. Neurosci.* 20, 491–496.

Ranalli, T. A., Tom, S., and Bambara, R. A. (2002), AP endonuclease 1 coordinates flap endonuclease 1 and DNA ligase I activity in long patch base excision repair, *J. Biol. Chem.* 277, 41715–41724.

Rapin, I. et al. (2000), Cockayne syndrome and xeroderma pigmentosum, *Neurology* 55, 1442–1449.

Reardon, J. T. et al. (1997), In vitro repair of oxidative DNA damage by human nucleotide excision repair system: possible explanation for neurodegeneration in xeroderma pigmentosum patients, *Proc. Natl. Acad. Sci. USA* 94, 9463–9468.

Reiter, R. J. (1995), Oxygen radical detoxification processes during aging: the functional importance of melatonin, *Aging* (Milano) 7, 340–351.

Richter, C., Park, J. W., and Ames, B. N. (1988), Normal oxidative damage to mitochondrial and nuclear DNA is extensive, *Proc. Natl. Acad. Sci. USA* 85, 6465–6467.

Rolig, R. L., and McKinnon, P. J. (2000), Linking DNA damage and neurodegeneration, *Trends Neurosci.* 23, 417–424.

Sanchez-Ramos, J., Overvik, Eva., Ames, B. (1994), A marker of oxyradical-mediated DNA damage (8-hydroxy-2'-deoxyguanosine) is increased in nigro-striatum of Parkinson's disease brain, *Neurodegeneration* 3, 197–204.

Schapira, A. H. et al. (1989), Mitochondrial complex I deficiency in Parkinson's disease, *Lancet* 1, 1269.

Schraufstatter, I. et al. (1988), Oxidant-induced DNA damage of target cells, *J. Clin. Invest.* 82, 1040–1050.

Sedelis, M. et al. (2000), MPTP susceptibility in the mouse: behavioral, neurochemical, and histological analysis of gender and strain differences, *Behav. Genet.* 30, 171–182.

Sekiguchi, M. (1996), MutT-related error avoidance mechanism for DNA synthesis, *Genes Cells* 1, 139–145.

Shelke, R. R., and Leeuwenburgh, C. (2003), Life-long calorie restriction (CR) increases expression of apoptosis repressor with a caspase recruitment domain (ARC) in the brain, *FASEB J.* 17, 494–496.

Simic, M. G., Bergtold, D. S., and Karam, L. R. (1989), Generation of oxy radicals in biosystems, *Mutat. Res.* 214, 3–12.

Singhal, R. K., Prasad, R., and Wilson, S. H. (1995), DNA polymerase beta conducts the gap-filling step in uracil-initiated base excision repair in a bovine testis nuclear extract, *J. Biol. Chem.* 270, 949–957.

Slupska, M. M. et al. (1999), Functional expression of hMYH, a human homolog of the Escherichia coli MutY protein, *J. Bacteriol.* 181, 6210–6213.

Sokoloff, L. (1960), The metabolism of the central nervous system in vivo, *Handbook of Physiology-Neurophysiology*, ed. J. Field, H. W. Magoun, and V. E. Hall, American Physiological Society, Washington, DC, 1843–1864.

Stedeford, T. et al. (2001), Comparison of base-excision repair capacity in proliferating and differentiated PC 12 cells following acute challenge with dieldrin, *Free Radic. Biol. Med.* 31, 1272–1278.

Stucki, M. et al. (1998), Mammalian base excision repair by DNA polymerases delta and epsilon, *Oncogene* 17, 835–843.

Sunesen, M. et al. (2002), Global genome repair of 8-oxoG in hamster cells requires a functional CSB gene product, *Oncogene* 21, 3571–3578.

Svejstrup, J. Q. (2002), Mechanisms of transcription-coupled DNA repair, *Nat. Rev. Mol. Cell Biol.* 3, 21–29.

Takao, M. et al. (1998), Mitochondrial targeting of human DNA glycosylases for repair of oxidative DNA damage, *Nucleic Acids Res.* 26, 2917–2922.

Timson, D. J., Singleton, M. R., and Wigley, D. B. (2000), DNA ligases in the repair and replication of DNA, *Mutat. Res.* 460, 301–318.

Tomkinson, A. E. et al. (1990), Mammalian DNA ligases. Catalytic domain and size of DNA ligase I, *J. Biol. Chem.* 265, 12611–12617.

Tomkinson, A. E., and Levin, D. S. (1997), Mammalian DNA ligases, *Bioessays* 19, 893–901.

Tuo, J. et al. (2002), The cockayne syndrome group B gene product is involved in cellular repair of 8-hydroxyadenine in DNA, *J. Biol. Chem.* 277, 30832–30837.

Venter, J. C. et al. (2001), The sequence of the human genome, *Science* 291, 1304–1351.

Vidal, A. E. et al. (2001), Mechanism of stimulation of the DNA glycosylase activity of hOGG1 by the major human AP endonuclease: bypass of the AP lyase activity step, *Nucleic Acids Res.* 29, 1285–1292.

Wang, J. Y., and Shum, A. Y. (2002), Hypoxia/reoxygenation induces cell injury via different mechanisms in cultured rat cortical neurons and glial cells, *Neurosci. Lett.* 322, 187–191.

Waters, T. R. et al. (1999), Human thymine DNA glycosylase binds to apurinic sites in DNA but is displaced by human apurinic endonuclease 1, *J. Biol. Chem.* 274, 67–74.

Wilson, D. M., 3rd, and Thompson, L. H. (1997), Life without DNA repair, *Proc. Natl. Acad. Sci. USA* 94, 12754–12757.

Wyatt, M. D. et al. (1999), 3-methyladenine DNA glycosylases: structure, function, and biological importance, *Bioessays* 21, 668–676.

Zharkov, D. O. et al. (2000), Substrate specificity and reaction mechanism of murine 8 oxoguanine-DNA glycosylase, *J. Biol. Chem.* 275, 28607–28617.

Zimmer, J., and Geneser, F. A. (1987), Difference in monoamine oxidase B activity between C57 black and albino NMRI mouse strains may explain differential effects of the neurotoxin MPTP, *Neurosci. Lett.* 78, 253–258.

Zuddas, A. et al. (1994), In brown Norway rats, MPP+ is accumulated in the nigrostriatal dopaminergic terminals but it is not neurotoxic: a model of natural resistance to MPTP toxicity, *Exp. Neurol.* 127, 54–61.

CHAPTER 8

Intracellular Signaling and Developmental Neurotoxicity

Prasada Rao S. Kodavanti

CONTENTS

8.1 INTRODUCTION

The effect of drugs and chemicals on the developing nervous system is an important and challenging public health problem. Damage to the developing nervous system by exposure to chemical agents can be accompanied by subtle to profound effects either on the structure or function of the nervous system, which can have lifelong consequences. An agent causing these effects is defined as a developmental neurotoxicant. It is important to note that the developmental phase at the time of exposure is a major determinant of the capacity of a chemical to produce developmental neurotoxicity.

The nervous system, which consists of the brain, the spinal cord, and a vast array of processes, coordinates the activities of the organ systems of the body and is responsible for communication with the surrounding environment, both internal and external. The smooth coordination of motor activity and cognitive functions such as speech, emotion, learning, and memory are dependent on the normal functioning of the nervous system. Many chemicals can alter the normal functioning of the nervous system, resulting in adverse health effects ranging from impairments in muscular movement, disruption of sensory functioning (hearing and vision) or memory loss. By altering the neurochemistry or physiology of the neurons during development, toxicants can impair the intracellular environment, which could lead to structural deficits including changes in the morphology of nerve cells and their subcellular elements.

A number of chemicals have proven to be more toxic to the developing nervous system than to the adult. For example, lead causes psychological deficits and interference with academic development in children of mothers who are asymptomatic (Davis et al., 1990). Likewise, methylmercury, which can induce congenital mental retardation and cerebral palsy in children, causes no or a few reversible neurological symptoms in mothers (Burbacher et al., 1990). Other examples include the special vulnerability of the developing brain to polychlorinated biphenyls (Tilson et al., 1990), ethanol (Driscoll et al., 1990), prescription drugs such as isotretinoin (Adams, 1990), and ionizing radiation (Schull et al., 1990). While the mechanisms of action for these agents have yet to be determined, the basis for the unique vulnerability of the nervous system may be based on developmental processes that regulate growth and development. Key processes in development include neurogenesis, where neurons proliferate followed by cell migration in which nascent neurons travel to their destinations. This stage is followed by differentiation, which includes neurite outgrowth (axonal and dendritic elongation) and arborization of dendritic spines. These processes overlap in time during development of the nervous system with other important developmental processes, including proliferation, migration, growth, and maturation (Figure 8.1). This chapter covers the sensitivity and chronology of nervous system development, the role of intracellular signaling in nervous system development and function, and finally, the role of intracellular signaling in the developmental neurotoxicity of environmental chemicals with particular emphasis on persistent organic pollutants. The details about the sensitivity of the developing brain and critical periods of brain development are discussed below.

8.2 DEVELOPMENT OF THE NERVOUS SYSTEM

Development of the nervous system is a complex program which involves the coordinated growth of axons and their targets in a timely manner. To study the effects

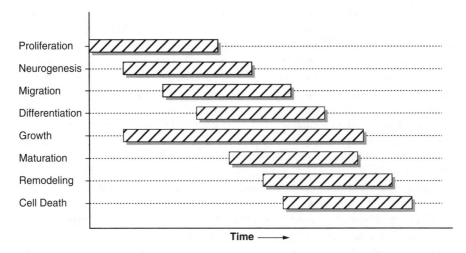

Figure 8.1   Temporal events in neuronal and synaptic development, which shows overlapping time periods of various processes. The time span is not specific for any brain region. Each brain region of the nervous system has its distinct progression through these processes. (Adapted from Gilbert, 1991.)

of toxicants on the nervous system, one must understand the developmental stages of the nervous system.

8.2.1   Chronology of Nervous System Development

It is known in rodents that some neurons are born very early (Pierce, 1973), even prior to closure of the neural tube, which occurs on embryonic day (E) 9 to 10 (Theiler, 1972). Neurons are produced throughout the remainder of gestation, but production levels are low when birth approaches. After birth, high rates of neuronal proliferation are seen in several brain areas, but neurogenesis is virtually complete by the end of the third week of postnatal life in the rodent brain. A striking feature of the genesis of many cell types is that production is completed in a very short time although long proliferative periods are associated with some medium-sized neurons. For example, mouse Purkinje cells of the cerebellum are produced by bursts of activity that last only 2 to 3 days (Miale and Sidman, 1961). This aspect is particularly important in understanding the neurotoxicity of chemicals where a brief insult during critical periods of development could cause substantial neuronal loss.

In many CNS regions, large neurons are formed first and small neurons follow. This timing was suggested to be important for laying down basic connections between neuronal groups. Cells that are widely separated in the adult brain are close together early in development (Sidman, 1969). As large neurons form basic connections between neuronal groups, smaller neurons are involved in local circuits. Among the parts of the nervous system, the spinal cord and brain stem are formed early. However, the ontogeny-phylogeny correlation may not apply for all the parts of the nervous system. For example, the archicortex forms later than the neocortex in all the species that have been studied (Rakic and Sidman, 1970). However, it is not true that the time of functional maturity of a system can be predicted from the time of origin of its neurons (Rakic and Sidman, 1970). There is less evidence available on the chronology of synaptogenesis (differentiation and maturation processes shown in Figure 8.1) than on the chronology of neurogenesis, but it seems possible that synaptogenesis may be better correlated with encephalization than neurogenesis. For example, some functions mediated by monkey prefrontal cortex do not mature until the third year of life (Alexander and Goldman, 1978), even though the cortical neurons and their target neurons form prenatally. In contrast, the firing pattern of Purkinje cells assumes many adult characteristics while neighboring neurons are still forming (Woodward et al., 1969).

In this chapter, the emphasis is on data from the mouse and rat, the most popular species in scientific research. Although no species is an ideal model for human brain development in terms of absolute time schedules, all the mammals studied have been shown to be similar in the sequence in which neurons form. The degree of functional maturity at birth differs from species to species, with humans somewhat more advanced than mouse and rat, but considerably less advanced than the monkey (Howard, 1973). Figure 8.2 shows the general outline of the bursts of proliferative activity that produce the normal nervous system in the mouse. Only the cerebral cortex, the hippocampus, and the cerebellum are shown in the figure in order to highlight the fact that development of brain regions occur at different times, some prenatal and some postnatal. The details on other brain regions can be found in the review article by Rodier (1980). In the cerebral cortex, the innermost layers of neurons are produced early starting at embryonic day 12 (E12). The time of origin of the cerebral cortex is later than the brain stem, but much earlier than the hippocampus and cerebellum. The pyriform and entorhinal cortex are the first regions to form, where proliferation ceases about E16. Some regions, such as the occipital cortex, start early but take longer to form (Angevine and Sidman, 1961). In the hippocampus, few neurons in the pyramidal cell layer arise as early as E11. The CA2 cell field is completed relatively quickly, between E13 and E16. Proliferation of granule cells in the dentate gyrus peaks perinatally, and the bulk of these neurons form after birth. In the cerebellum, the Purkinje cells are among the first neurons to proliferate on E11–13 (Andreoli et al., 1974). After Purkinje cell production ceases, cells migrate from the ventricular germinal zone and spread over the surface of the presumptive cerebellum, forming the external germinal layer. This special proliferative zone grows to a maximal thickness during the second postnatal week, and then gradually regresses. From it arise the cells of the molecular and granular layers (Miale and Sidman, 1961). Most molecular-layer neurons form during the first week of life while granule cell neurons form just before birth to at least postnatal day 15.

8.2.2 Sensitivity of the Nervous System during Development

The complexity of the nervous system contributes to its vulnerability to toxicant insult. Unlike other organs in the body, the brain has several regions with distinct functional characteristics. The nervous system is also composed of a wide variety of neuron and glial cell types. Neurons can be classified on the basis of their

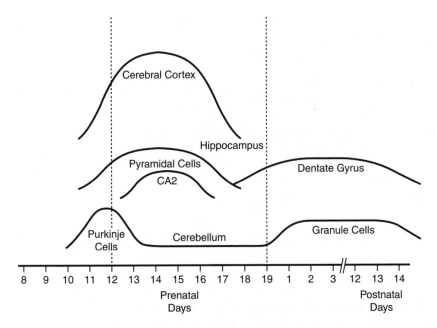

Figure 8.2 Chronology of mouse nervous system development. Examples of a few brain regions are shown here. Please note the different time span for development of cell types within a brain region. Also, the overlap of different processes during development of the neuron. (Adapted from Rodier, 1980.)

anatomical location (e.g., hippocampal, striatal), primary neurotransmitter (e.g., cholinergic, dopaminergic), cellular structure (e.g., granule, pyramidal), or function (e.g., neuroendocrine, sensory). Astrocytes, oligodendrocytes, and microglia have been identified as three classes of neuroglia, which are responsible for a broad range of functions including neuronal support and maintenance, myelin production, and phagocytosis. Preservation of the normal relationship between neurons and glia is critical for the maintenance of normal nervous system function (Abbott et al., 1992; LoPachin and Aschner, 1999). Neurotoxicants can disrupt this relationship by selectively damaging either cell population.

The nervous system has a very high metabolic rate that is almost entirely dependent on aerobic glucose metabolism. In order to support this high metabolic demand, the brain receives approximately 15% of the total cardiac output and accounts for 20% of oxygen consumed by the body, even though the brain weighs only 1.5 to 2% of the total body weight (Heiss, 1981). Thus, the nervous system is very sensitive to toxicants that disrupt mitochondrial function and energy metabolism (Nichlas et al., 1992). The high metabolic rate, low to moderate levels of anti-oxidants, and high content of polyunsaturated fatty acids predispose the brain to oxidative damage (Evans, 1993). The nervous system needs a lot of energy during the development. In the rat brain, maturation of oxidative metabolism in mitochondria is initiated in late gestation, but acceleration in mitochondrial respiration occurs immediately after birth in order to maintain high-energy phosphate levels, which is crucial for the successful outcome of the newborn (Nakai et al., 2000).

One of the most crucial characteristics of the developing nervous system for its vulnerability is the immaturity of the blood-brain barrier (Risau and Wolburg, 1990). Once drugs or environmental chemicals cross the placenta, they are readily accessible to the embryonic and fetal brain due to their inability to exclude these agents. Since the placenta is permissive to the transport of many molecules, agents in maternal circulation will reach the developing brain in concentrations at least equal to those attained in the whole embryo or fetus. Once the chemical is in the brain of the fetus, it can have effects on the ontogeny of a number of critical processes described below. Thus, the developing nervous system may be more vulnerable for disruption by chemicals than the mature brain because exclusionary processes are not yet developed at the time when cells are rapidly proliferating, migrating, differentiating, and connecting.

8.3 INTRACELLULAR SIGNALING MECHANISMS

Intracellular signaling or signal transduction can be defined as a mechanism by which extracellular signals are transferred to the cytosol and nucleus of the cell. Although 17 intracellular signaling pathways have been identified based on their ligands (NRC, 2000), only those signaling pathways relevant to the developing mammalian nervous system are discussed here. These signals are not only essential for the function of the nervous system; they also play a key role in nervous system development. Any interference with these processes would have the potential to have profound effects on the function of neurons as well as on their development. Growth factors, neurotransmitters, and hormones serve as first messengers to transfer information from one cell to another by interacting with specific cell membrane receptors. This interaction leads to activation or inhibition of specific enzymes or opening of ion channels, which results in changes in intracellular metabolism, leading to a variety of effects including activation of protein kinases and transcription factors. These intracellular pathways can be activated by totally different receptors, and there is significant cross talk between the receptor signals, so they can control and modulate each other.

The known systems that link membrane receptors to second messenger systems include G-protein coupled receptors, growth factors with tyrosine kinase activity, and neurotransmitter receptors. Several GTP-binding proteins have been identified that couple receptors to intracellular signaling processes such as phospholipid hydrolysis and cyclic nucleotide metabolism (Simon et al., 1991). Stimulation or inhibition of adenylate cyclase activity involves binding of GTP to the receptors, and the end result is a change in the intracellular levels of cyclic AMP, which in turn binds to the regulatory subunit of protein kinase A (Krebs, 1989). Hydrolysis of membrane phosphatidylinositol or phosphatidylcholine by phospholipases C or D generate intracellular second messengers such as inositol 1,4,5-trisphosphate (IP3), phosphatidic acid, and diacylglycerol (Fisher et al., 1992; Nishizuka, 1995; Exton, 1996; Liscovitch, 1996; English, 1996; Liu, 1996). Diacylglycerol activates protein kinase C (PKC), while IP3 releases calcium from intracellular stores such as the endoplasmic reticulum. Generation of arachidonic acid (AA), which can also activate PKC (Khan et al., 1995), by phospholipase A_2 and AA metabolism to several products with different biological activities, is another relevant signal transduction process (Bonventre, 1996). Growth factor receptors such as tyrosine kinases, upon activation, dimerize and autophosphorylate to initiate several intracellular signal transduction pathways (Malarkey et al., 1995). An important family of kinases, the mitogen-activated protein kinases (MAP kinases), can be activated by both receptor kinases and G-protein coupled receptors (Ferrell, 1996). A common pathway starting from the receptor tyrosine kinase involves a small GTP-binding protein (p21ras) that activates Raf-1, which activates MAP kinase kinase (MEK), which in turn phosphorylates and activates MAP kinase. Other second messengers include nitric oxide (NO), produced by activation of nitric oxide synthase (NOS), which converts L-arginine to L-citrulline and NO (Zhang and Snyder, 1995). NO can activate guanylate cyclase, which generates cyclic GMP, which in turn activates protein kinase G (Wang and Robinson, 1997). All these signaling processes play key roles in the normal functioning of the nervous system as well as in the development of the nervous system.

8.3.1   Role in Nervous System Function

In terms of developmental neurotoxicity, motor coordination and learning/memory are two functional aspects that are well studied following chemical challenge. The second messengers, generated from the intracellular signaling pathway, play a major role in both of these functions (Serrano et al., 1994; Tanaka and Nishizuka, 1994; Kaul et al., 1996; Chen et al., 1997; Atkins et al., 1998; Sukhotina et al., 1999; Berman et al., 2000; Godinho et al., 2002). Two key second messengers such as calcium and PKC are discussed here as examples of the role of intracellular signaling in the biological functions of the brain.

Calcium plays an integral part in the function of the nervous system. The physiological levels of intracellular free calcium (100 to 300 nM) are maintained by the effective operation of calcium pumps located in plasma membrane and sequestration into intracellular organelles such as mitochondria and endoplasmic reticulum. Perturbations in calcium homeostasis can lead to altered functions such as neurotransmitter release (Katz, 1969; Smith and Augustine, 1988), neuronal excitability (Marty, 1989), integration of electrical signals (Llinas, 1988; Marty, 1989), synaptic plasticity of various types (Malenka et al., 1989; Fitzjohn and Collingridge, 2002; Levy et al., 2003), gene expression (Szekely et al., 1990), metabolism (McCormack et al., 1990), or programmed cell death (Chalfie and Wolinsky, 1990). Additional information on the role of calcium in nervous system function is available (Carafoli, 1987; Gibson and Peterson, 1987; Siesjo, 1990; Miller, 1991).

PKC signaling pathway has been implicated in the modulation of motor behavior as well as learning/memory (Serrano et al., 1994; Tanaka and Nishizuka, 1994; Chen et al., 1997). Injection of PKC inhibitor H7 into the dopaminergic region blocks the acute response to cocaine (Steketee, 1993). Inhibitors of PKC also block the development of long-term potentiation (LTP, a physiological form of synaptic plasticity and learning) when administered in the first two hours after induction (Jerusalinsky et al., 1994). PKC is a phospholipid-dependent enzyme consisting of at least 12 isozymes, which include conventional PKCs (PKC-alpha, beta, gamma) that require diacylglycerol, phosphatidylserine, and calcium; novel PKCs (PKC-delta, epsilon, neu, and theta) which are not dependent on calcium but dependent on diacylglycerol; and atypical PKCs (PKC-zeta, iota, and lambda) whose regulation require neither diaylglycerol nor calcium. PKC isozymes are differentially distributed in brain and their roles are isozyme-specific (Majewski and Iannazzo, 1998). Specific PKC isozymes appear to be involved in induction of LTP. For example, PKC-gamma is associated with lead-induced impairment of synaptic plasticity, learning and memory in rats (Nihei et al., 2001). Transgenic mice lacking PKC-gamma also exhibit impairment of LTP and spatial and contextual learning (Abeliovich et al., 1993). An activation and relocation of PKC following learning processes was also observed during the acquisition of nictitating membrane conditioning in the rabbit (Sunayashi-Kusuzaki et al., 1993). Both calcium-dependent and -independent PKCs have also been shown to be changed in the brains of patients with Alzheimer's disease (Matsushima et al., 1996).

8.3.2   Role in Nervous System Development

The key role played by extracellular factors such as hormones, neurotransmitters, and growth factors in the development of the nervous system has been well established. (For reviews, see Lanier et al., 1976; Mattson, 1988; McDonald and Johnston, 1990; Costa, 1993, 1994.) The binding of these extracellular factors to cell surface receptors triggers the onset of intracellular signaling leading to the production of several second messengers including phosphoinositides and cyclic nucleotides. Phosphoinositide metabolism, stimulated by neurotransmitters in general, changes with age (Schoeppe and Rutledge, 1985; Balduini et al., 1991). However, the ability of muscarinic, glutamatergic, and serotonergic receptors to stimulate phosphoinositide metabolism is enhanced in the immature rat (Balduini et al., 1987; Claustre et al., 1988; Palmer et al., 1990). The developmental profile is different for each of these neurotransmitters and among brain regions (Palmer et al., 1990; Balduini et al., 1991). The enhanced stimulation of phosphoinositide metabolism coincides with the so-called "brain growth spurt" during an early postnatal period (days 1 to 15) in the rat suggesting the role of phosphoinositides in development. This brain growth spurt is characterized by dendritic arborization, establishment of synaptic contacts, and astroglial proliferation (Rodier, 1980).

There is also ample evidence that phosphoinositide metabolism, leading to elevations in intracellular free calcium and activation of PKC, plays multiple important roles in brain development (Murphy et al., 1987; Girard and Kuo, 1990). Activation of PKC by phorbol esters has been shown to stimulate astrocyte proliferation in primary cultures (Murphy et al., 1987) as well as enhance differentiation of astrocytes and oligodendrocytes in culture (Honegger, 1986; Yong et al., 1988). Proliferation of Schwann cells is also accompanied by enhanced phosphoinositide metabolism (Saunders and De Vries, 1988). Inhibitors of PKC, such as calphostin C, reduced neurite initiation and axon branching without significantly affecting the number of dendrites per neuron (Cabell and Audesirk, 1993). The peculiar developmental pattern of PKC

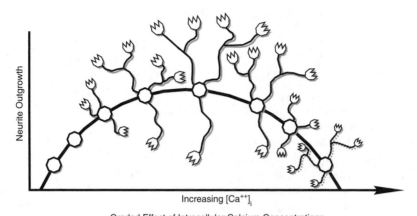

Graded Effect of Intracellular Calcium Concentrations

At LOW [Ca^{++}]$_i$: Neurons fail to survive
PHYSIOLOGICAL [Ca^{++}]$_i$: Promote successive increases in outgrowth
HIGHER [Ca^{++}]$_i$: Outgrowth decreases, and pruning of existing neurons
VERY HIGH [Ca^{++}]$_i$: Neuronal cell death

Figure 8.3   Role of signal transduction pathways (e.g., Ca^{++}) in the normal growth of neurons. Cells need to maintain intracellular free calcium at physiological concentration in order to function and grow normally. When intracellular levels of calcium were lower, neurons fail to survive. At higher concentrations, neuronal outgrowth decreases and pruning of existing neurons occur. At very high concentrations, neuronal death occurs. (Adapted from Kater and Mills, 1991.)

isozymes in the brain (Hashimoto et al., 1988; Sposi et al., 1989; Shearman et al., 1991) suggests that PKC plays a key role in various aspects of brain development. The important role of calcium in the viability of neurons and its outgrowth has been well established (Figure 8.3). Intraneuronal free calcium levels should be in the physiological range between 100 and 300 nM to promote successive increases in outgrowth. At levels below 100 nM, neurons fail to survive and are associated with no growth. At concentrations slightly higher than 300 nM, neurite outgrowth decreases and pruning of existing neurons occurs. As seen in Figure 8.3, at very high concentrations, the death of neurons occurs (Kater and Mills, 1991). Calcium has been shown to induce differentiation of neuroblastoma cells and neurite extension in PC12 (Pheochromocytoma) cells (Traynor, 1984; Reboulleau, 1986).

Other important second messengers in signal transduction are the cyclic nucleotides. The levels of cyclic AMP increase with age in the rat brain (Schmidt et al., 1980) without any significant changes in adenylate cyclase activity (Keshles and Levitzki, 1984). However, agonist-stimulated adenylate cyclase increases significantly with age (Ma et al., 1991). Several studies have shown that cyclic AMP can either enhance or inhibit neurite outgrowth, depending on the neuronal type (Mattson, 1988). In embryonic rat cortical cells and hippocampal pyramidal neurons, increased cyclic AMP levels were associated with enhanced neurite extension (Shapiro, 1973; Mattson et al., 1988). On the other hand, cyclic AMP inhibits neurite elongation and growth cone motility in Helisoma neurons (Mattson et al., 1988).

8.3.3   Regulation of Downstream Molecular Events

The downstream events following production of intracellular messengers from the extracellular stimuli culminate in the activation of nuclear transcription factors (Edwards, 1994). It has been shown that alterations in calcium homeostasis and PKC resulted in altered gene expression of transcription factors, neurotrophic factors or neuropeptides which were correlated with changes in structural proteins implicated in synaptic plasticity (Bading, 1999; Mayer et al., 2002). While the expression of immediate early genes (e.g., c-fos, c-jun) is very low in quiescent cells, these genes are rapidly induced at the transcriptional level (Sheng and Greenberg, 1990). The activation of these early genes can result in the activation of target genes, which may code for a variety of structural and functional proteins (Morgan and Curran, 1989; Sheng and Greenberg, 1990; Hall, 1990). A number of proteins involved in the development and normal function of the nervous system have been shown to be phosphorylated by events in the intracellular signaling process. These proteins include microtubule-associated proteins (MAPs) expressed mainly in nerve cell bodies, dendrites, and axons; neural cell adhesion molecules (NCAM) involved in neurogenesis; synaptic proteins such as synapsin, synaptobrevin, synaptotagmin; growth associated protein (GAP-43) expressed in growth cones; and myelin basic proteins involved in myelination. The details about transcription factors and their role in neurotoxicity can be found in other chapters in this book.

8.4 DEVELOPMENTAL NEUROTOXICANTS AND INTRACELLULAR SIGNALING

In recent years, investigators have considered a number of signal transduction pathways as potential targets for neurotoxicants including metals, alcohols, and persistent environmental pollutants. Neurotoxicants may alter any step beginning from the extracellular stimulus all the way to the transcriptional level leading to alterations in the structure or function of the nervous system. Intracellular signaling could be a potential mechanism by which chemicals alter nervous system function directly or through alterations in nervous system morphology. Thus, brain function and morphology may be affected by neurotoxicants through changes in intracellular signaling at critical phases during development. Several chemicals, including alcohols, metals, and psychoactive drugs have been identified as developmental neurotoxicants in humans and animals. Anderson et al. (2000) identified 9 pesticides, 14 solvents, 10 metals or organometal compounds, and 6 other industrial chemicals as developmental neurotoxicants. A similar number of chemicals were identified as developmental neurotoxicants by Goldey et al. (1995). Table 8.1 provides a partial list of chemicals, which includes physical agents such as X-irradiation. Some of the known developmental neurotoxicants, their effects on intracellular signaling and their observed effects in animals and humans are presented in Table 8.2. Although there are a number of developmental neurotoxicants belonging to different categories (see Table 8.1), this chapter focuses on a particular group of pollutants known as polychlorinated biphenyls (PCBs). Information on other pollutants is available in

Table 8.1   Partial List of Chemicals Believed to Be Developmental Neurotoxicants

Category	Chemical/Drug
Alcohols	Methanol, ethanol
Antimitotics	Azacytidine, methylazoxy-methanol acetate
Organochlorine insecticides	DDT, chlordecone, endrin
Organophosphates	Methyl parathion, chlorpyrifos, fenitrothion
Herbicides	2,4-D/2,4,5-T
Fumigants	Ethylene bromide
Metals	Lead, methylmercury
Organotins	Triethyltin, trimethyltin
Polyhalogenated hydrocarbons	PCBs, PBBs, PBDEs (?)
Psychoactive drugs	Cocaine, methadone
Therapeutic agents	Diphenylhydantoin
Solvents	Carbon disulfide, toluene, trichloroethylene, xylene
Vitamins	Vitamin A
Physical agents	X-irradiation

Note: These chemicals were reported to cause adverse effects in offspring following developmental (pre- and/or post-natal exposure). PCBs: polychlorinated biphenyls; PBBs: polybrominated biphenyls; PBDEs: polybrominated diphenyl ethers; DDT: dichlorodiphenyl trichloroethane.

Adapted from Goldey et al. (1995) and Anderson et al. (2000).

the literature. (For reviews, see Burbacher et al., 1990; Davis et al., 1990; Driscoll et al., 1990; Vorhees, 1992; Goldey et al., 1995; Eriksson, 1997; Anderson et al., 2000; Costa, 1998a; Costa et al., 2001.) Metals and alcohols are briefly discussed here.

8.4.1   Metals

Among metals, methylmercury and lead are well studied for their effects on the developing nervous system. These metals have shown developmental neurotoxic effects both in humans and in animals (Burbacher et al., 1990; Davis et al., 1990; Table 8.2). Methylmercury induces neuropathological and neurobehavioral effects, which exhibit striking comparability across a number of species, including humans. The most striking effects from lead exposure are those in the area of cognitive development, where multiple studies in humans, monkeys, and rats have shown performance deficits on tests of intelligence and learning (Davis et al., 1990). Methylmercury and lead have been shown to affect several signaling pathways including calcium levels (Pounds, 1984; Goldstein, 1993; Oyama et al., 1994; Mundy and Freudenrich, 2000; Limke et al., 2003), calcium channels (Audesirk, 1993; Sirois and Atchison, 2000), and PKC (Long et al., 1994; Rajanna et al., 1995; Haykal-Coates et al., 1998; Reinholz et al., 1999; Nihei et al., 2001). Developmental exposure to methylmercury has been shown to affect neuronal migration and neurogenesis in the cerebral cortex (Kakita et al., 2002; Faustman et al., 2002). In PC12 cells, methylmercury has shown to decrease NGF-induced TrkA autophosphorylation and neurite outgrowth (Parran et al., 2003). Likewise, lead exposure resulted in inhibition of neurite development in hippocampal neurons (Kern and Audesirk, 1995).

Table 8.2 Interaction of Some Pharmacological Agents and Known/Suspected
Neurotoxicants in Animals/Humans on Intracellular Signaling Pathways

Chemical Exposure	Intracellular Signal Transduction Pathway (e.g., Calcium, PKC, Phosphoinositol, etc.)	Neurotoxicity— Animals (e.g., Cognitive Function, LTP or Structural Change)	Neurotoxicity— Humans (e.g., Learning and Memory)
PKC inhibitor[a]	YES	YES	n/a
PKC knock-out[b]	YES	YES	n/a
MAM[c,d]	YES	YES	n/a
Lead[e]	YES	YES	YES
Aluminum[f]	YES	YES	YES
Methylmercury[g]	YES	YES	YES
Organotins[h]	YES	YES	YES
Ethanol[i]	YES	YES	YES
Cyanide[j]	YES	YES	YES
DDT[k]	YES	YES	YES
PCBs[l]	YES	YES	YES
PBDEs[m]	YES?	YES?	?

Note: n/a: not applicable; ? indicates very limited information or no available information.

[a] Cabell and Audesirk, 1993; Steketee, 1993; Jerusalinsky et al., 1994.
[b] Abeliovich et al., 1993.
[c] Di Luca et al., 1994; Goldey et al., 1994; Caputi et al., 1996.
[d] MAM, methylazoxy-methanol acetate, a potent alkylating antimitotic agent, has been shown to alter PKC and brain morphometry, and to cause cognitive dysfunction during developmental exposure.
[e] Davis et al., 1990; Audesirk, 1993; Long et al., 1994; Kern and Audesirk, 1995.
[f] Pendlebury et al., 1987; White et al., 1992; Mundy et al., 1997.
[g] Burbacher et al., 1990; Haykal-Coates et al., 1998; Kakita et al., 2002.
[h] Besser et al., 1987; Walsh and DeHaven, 1988; Pavlakovic et al., 1995.
[i] Streissguth et al., 1980; Spohr et al., 1993; Costa, 1998b.
[j] Maduh et al., 1990; Yamamoto, 1990; Kanthasamy et al., 1994.
[k] Kodavanti et al., 1996b; Eriksson, 1997; Narahashi, 2000; Dorner and Plagemann, 2002.
[l] Schantz et al., 1995; Gilbert and Crofton, 1999; Patandin et al., 1999; Kodavanti and Tilson, 1997; Yang et al., 2003.
[m] Eriksson, 1997; Kodavanti and Derr-Yellin, 2002.

8.4.2 Alcohols

Ethanol is a well-studied alcohol for its effects on the developing nervous system (see Chapter 10). Clinically, ethanol-induced teratogenecity has been characterized as the fetal alcohol syndrome, which includes CNS effects such as mild to moderate mental deficiency, attention deficit-hyperactivity disorder, learning deficiency, altered reaction times, altered reflexes, and delayed development (Streissguth et al., 1980; Spohr et al., 1993). Ethanol has been shown to affect several signal transduction pathways in different model systems (Costa, 1998b). Both cyclic AMP metabolism and phospholipid/PKC pathway have been affected by exposure to ethanol (Hoek and Rubin, 1990; Hoffman and Tabakoff, 1990). *In vivo* administration of ethanol to rats during the brain growth spurt (postnatal days four through ten) caused significant microencephaly without affecting pup's body weight and a significant decrease in muscarinic receptor-stimulated phosphoinositide metabolism (Balduini and Costa, 1989). Two other alcohols, t-butanol and n-propanol, which have been shown to cause microencephaly in the rat when administered during the brain growth spurt, were also potent inhibitors of muscarinic-receptor-stimulated phosphoinositide metabolism (Grant and Samson, 1982, 1984). On the other hand, methanol had only a minimal effect on these events (Candura et al., 1991). With regard to cyclic AMP metabolism, the interaction of ethanol appears to involve in particular to Gs, which belongs to guanine nucleotide binding proteins (Hoffman and Tabakoff, 1990). Ethanol exposure during development has also been associated with neuronal loss, particularly in the hippocampus and cerebellum (Hoffman and Tabakoff, 1996). The literature indicates that ethanol interacts with intracellular signaling processes in different model systems, and some of these interactions may be related to its effects on the developing nervous system.

8.4.3 Persistent Organic Pollutants

PCBs belong to a group of persistent organic pollutants called polyhalogenated aromatic hydrocarbon family. This family includes other pollutants such as poly-chlorinated dibenzo-*p*-dioxins, polychlorinated dibenzofurans, polychlorinated diphenyl ethers, and polybrominated diphenyl ethers (Figure 8.4). PCBs exist as mixtures containing up to 209 possible congeners and are ubiquitous environmental contaminants resulting from intensive industrial use and inadequate disposal over

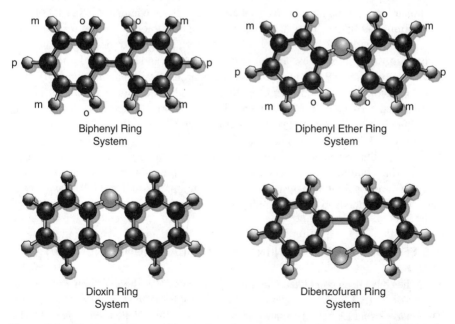

Biphenyl Ring
System

Diphenyl Ether Ring
System

Dioxin Ring
System

Dibenzofuran Ring
System

Figure 8.4   Structural features of a number of persistent organic pollutants, believed to be developmental neurotoxicants in humans and animals. These chemicals have shown to be persistent in the environment and bioaccumulate in the biological systems. Note the structural similarities in these environmental contaminants.

past decades (Erickson, 1986). PCB mixtures as well as congeners possess a surprising array of biological activity leading to toxicity (WHO, 1993). It is known that some PCBs and other halogenated hydrocarbons such as 2,3,7,8-tetrachlorodibenzo-*p*-dioxin (TCDD) produce their biological effects through a receptor-mediated response by binding to the cytosolic aryl hydrocarbon (Ah) receptor followed by induction of a number of genes (Safe, 1994; Okey et al., 1994). It has been proposed that non-*ortho* substitutions on the biphenyl ring (lateral substitutions at the *meta*- and *para*- positions) promote coplanarity and are associated with the TCDD-like toxic effects of certain PCB congeners (McKinney and Waller, 1994). Ah receptor involvement is associated with reproductive, immunologic, teratogenic, and carcinogenic effects of PCBs (Safe, 1994; Kafafi et al., 1993); however, other studies indicate that neurotoxic effects of PCBs might not be mediated through the Ah receptor (Schantz, 1996; Seegal, 1997; Tilson and Kodavanti, 1997, 1998).

There is now both epidemiological and experimental evidence that developmental exposure to PCBs causes cognitive deficits in humans (Jacobson and Jacobson, 1996; Patandin et al., 1999) and animals (Schantz et al., 1995; Niemi et al., 1998); however, the underlying cellular or molecular mechanisms are not known (Kodavanti and Tilson, 1997; Tilson and Kodavanti, 1997). It has been hypothesized that altered intracellular signaling/second messenger homeostasis by PCBs may be associated with these effects. This hypothesis was based on the following reasons: (1) the most significant neurotoxic effects of PCBs seen in humans are learning and memory deficits (Jacobson and Jacobson, 1996; Patandin et al., 1999); (2) laboratory studies indicate that PCBs inhibit long-term potentiation (LTP, a form of synaptic plasticity) and impair learning/memory *in vivo* (Schantz et al., 1995; Niemi et al., 1998; Gilbert and Crofton, 1999); (3) LTP is often described as a physiological model for neuronal development, learning, and memory (Lynch, 1998); (4) second messengers such as calcium, inositol phosphates, arachidonic acid (AA), and nitric oxide have been shown to modulate LTP and play key roles in neuronal development (Lynch, 1998); and (5) epidemiological studies indicate that infants born to mothers who consumed contaminated cooking oil in Yusho, Japan showed abnormal calcification of the skull indicating that PCBs might interfere with calcium metabolism (Yamashita and Hayashi, 1985). A number of studies have been conducted both *in vitro* and *in vivo* to understand the effects of PCBs on second messenger systems. *In vitro* experiments using neuronal cultures as well as brain homogenate preparations and *in vivo* studies involving developmental exposure to a PCB mixture, Aroclor 1254, have been conducted to understand the PCB-induced neurotoxicity.

8.5 EFFECTS OF PCBs ON INTRACELLULAR SIGNALING *IN VITRO*

Studies have been conducted with prototypic *ortho* (2,2'-dichlorobiphenyl, DCB) and non-*ortho* (3,3',4,4',5-pentachlorobiphenyl; PeCB or 4,4'-DCB) PCBs followed by extensive structure-activity relationships with more than 35 PCB congeners. PCB-induced alterations in intracellular free Ca^{2+} ($[Ca^{2+}]_i$) with a fluorescent dye (Fluo-3AM) were reported for the first time; the *ortho*-substituted 2,2'-DCB was more effective than the non-*ortho*-substituted 3,3',4,4',5-PeCB (Kodavanti et al., 1993).

Table 8.3   *In Vitro* **Effects of PCBs and Related Chemicals on Intracellular Signaling Processes in Neuronal Cultures and Brain Homogenate Preparations**

Intracellular Signaling Process	Lowest Concentration with a Significant Effect			
	Ortho-PCB	Non-Ortho-PCB	PCDE	PBDE
Calcium homeostasis:				
Free calcium levels	25 μM	NEO	—	—
Calcium buffering	5 μM	100 μM	3 μM	1–3 µg/ml[a]
Calcium extrusion	10 μM	NEO	—	—
Inositol phosphates:				
Basal PI metabolism	100 μM	NEO	—	—
Carb-stimulated PI	30 μM	NEO	—	—
Arachidonic acid release	3 μM	NEO	—	10 µg/ml
PKC translocation:				
^{3}H-PDBu binding	30 μM	NEO	30 μM	3–10 µg/ml[a]
PKC Western blots (α and ε)	25 μM	NEO	—	—
Nitric oxide synthase:				—
Cytosolic NOS	10 μM	NEO		—
Membrane NOS	10 μM	NEO		—
Neurite outgrowth in PC12 cells	5 µg/ml[b]			

Note: For details, see Kodavanti and Tilson, 1997 and 2000; Tilson and Kodavanti, 1997; Kodavanti and Derr-Yellin, 2002. *Ortho*-PCB = 2,2′-dichlorobiphenyl; non-*ortho*-PCB = 3,3′,4,4′,5-pentachlorobiphenyl or 4,4′-dichlorobiphenyl. NEO = no effect observed even at 100 µM.

[a] Unpublished data.
[b] Angus and Contreras (1995) with Aroclor 1254 mixture.

The increase in $[Ca^{2+}]_i$ was slow, and a steady rise was observed with time (Kodavanti et al., 1993). The follow up studies confirmed these observations in cerebellar granule neurons (Carpenter et al., 1997; Mundy et al., 1999; Bemis and Seegal, 2000), cortical neurons (Inglefield and Shafer, 2000; Inglefield et al., 2002), and in human granulocytes (Voie and Fonnum, 1998). Studies characterizing the mechanisms of $[Ca^{2+}]_i$ increase indicated that 2,2′-DCB was an inhibitor of $^{45}Ca^{2+}$-uptake by mitochondria and microsomes with IC_{50} (concentration which inhibits control activity by 50%) values of 6–8 µM. 3,3′,4,4′,5-PeCB inhibited Ca^{2+}-sequestration, but the effects were much less than those produced by equivalent concentrations of 2,2′-DCB. As seen in Table 8.3, synaptosomal Ca^{2+}-ATPase, involved in Ca^{2+}-extrusion process, was only inhibited by 2,2′-DCB, but not by 3,3′,4,4′,5-PeCB (Kodavanti et al., 1993). Further structure-activity relationship (SAR) studies indicated that congeners that are noncoplanar inhibited $^{45}Ca^{2+}$-uptake by microsomes and mitochondria while coplanar congeners did not (Kodavanti et al., 1996a).

The disruption of Ca^{2+}-homeostasis may have a significant effect on other signal transduction pathways (e.g., inositol phosphate [IP] and AA second messengers) regulated or modulated by Ca^{2+}. The congener 2,2′-DCB, but not 3,3′,4,4′,5-PeCB, affected basal and carbachol (CB)-stimulated IP accumulation in cerebellar granule cells (Kodavanti et al., 1994). Further studies indicated that any modulation of CB-stimulated IP accumulation is due to Ca^{2+}-overload, but not due to activation of PKC activity (Kodavanti et al., 1994; Shafer et al., 1996). AA is released intracellularly following activation of membrane phospholipases, and AA is an important second messenger in releasing Ca^{2+} from endoplasmic reticulum (Striggow and Ehrlich, 1997). Aroclor 1254 (a commercial mixture of PCBs) and 2,2'-DCB increased [3H]-AA release in cerebellar granule cells while 4,4-DCB did not (Kodavanti and Derr-Yellin, 1999); this is in agreement with previous structure-activity relationship studies on Ca^{2+} buffering and PKC translocation (Kodavanti and Tilson, 1997). A similar increase in [3H]-AA was observed with structurally similar chemicals such as PBDE mixtures (Kodavanti and Derr-Yellin, 2002). Further studies indicated that the [3H]-AA release caused by these chemicals could be due to activation of both Ca^{2+}-dependent and -independent phospholipase A_2 (PLA_2).

One of the downstream effects of perturbed Ca^{2+}-homeostasis is translocation of PKC from the cytosol to the membrane where it is activated (Trilivas and Brown, 1989). [3H]-Phorbol ester ([3H]-PDBu) binding has been used as an indicator of PKC translocation. The congener 2,2'-DCB increased [3H]-PDBu binding in a concentration-dependent manner in cerebellar granule cells, while 3,3',4,4',5-PeCB had no effect in concentrations up to 100 μM (Table 8.3). The effect of 2,2'-DCB was time-dependent and also dependent on the presence of external Ca^{2+} in the medium. Sphingosine, a PKC translocation blocker, prevented 2,2'-DCB-induced increases in [3H]-PDBu binding (Kodavanti and Tilson, 2000). Experiments with several pharmacological agents revealed that the effects are additive with glutamate, and none of the channel (glutamate, calcium, and sodium) antagonists blocked the response of 2,2'-DCB (Kodavanti et al., 1994). Immunoblots of PKC-alpha and epsilon indicated that non-coplanar ortho-PCB decreased the cytosolic form and increased the membrane form significantly at 25 μM (Yang and Kodavanti, 2001). Subsequent structure-activity relationship (SAR) studies indicated that congeners that are noncoplanar increased PKC translocation while coplanar congeners did not (for review, see Kodavanti and Tilson, 1997 and 2000; Fischer et al., 1998). This was further strengthened by observations with structurally similar chemicals such as polychlorinated diphenyl ethers (Kodavanti et al., 1996b; Table 8.3). Nitric oxide (NO), which is produced by NOS, is a gaseous neurotransmitter. NO has an important role as a retrograde messenger in LTP, learning and memory processes, and endocrine function (Schuman and Madison, 1994; McCann et al., 1998). The congener 2,2'-DCB, but not 4,4'-DCB, inhibited both cytosolic (nNOS) and membrane (eNOS) forms of NOS (Sharma and Kodavanti, 2002).

These *in vitro* studies clearly demonstrated that second messenger systems, involved in the development of the nervous system, LTP, and learning and memory, are sensitive targets for the *ortho*-substituted PCBs and related chemicals. Figure 8.5 illustrates the intracellular signaling events affected by these chemicals (ortho-PCBs and commercial PCB mixtures) at low micromolar concentrations and shorter exposure periods, where cytotoxicity is not evident. These signaling pathways include calcium homeostasis and PKC translocation. The rise of intracellular free Ca^{2+} is slow, but steady following exposure. This free Ca^{2+} rise could be due to increased calcium influx, inhibited Ca^{2+} buffering mechanisms, and/or calcium release from intracellular stores by the products of membrane phospholipases. This increase in

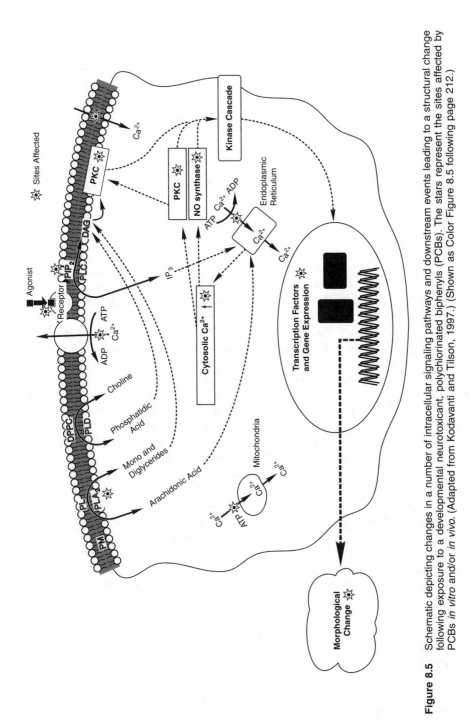

Figure 8.5 Schematic depicting changes in a number of intracellular signaling pathways and downstream events leading to a structural change following exposure to a developmental neurotoxicant, polychlorinated biphenyls (PCBs). The stars represent the sites affected by PCBs *in vitro* and/or *in vivo*. (Adapted from Kodavanti and Tilson, 1997.) (Shown as Color Figure 8.5 following page 212.)

free Ca^{2+} could cause translocation of PKC. The coplanar non-*ortho*-PCBs have marginal effects on calcium homeostasis and no effects on PKC translocation. Literature reports indicate that at slightly higher concentrations, commercial PCB mixtures (Aroclors 1221 and 1254) have been able to alter neurite outgrowth in PC12 cells (Angus and Contreras, 1995) and in hypothalamic cells (Gore et al., 2002). The possible mode of action for this structural change could be due to changes in intracellular signaling by these chemicals.

8.6 EFFECTS OF PCBs ON INTRACELLULAR SIGNALING *IN VIVO*

In vivo effects of PCBs have been studied with a commercial PCB mixture, Aroclor 1254, given orally from gestational day 6 through postnatal day 21. Arolcor 1254 treatment did not alter maternal body weight or percent mortality caused a small and transient decrease in body weight gain of offspring. Both calcium homeostasis and PKC activities were significantly affected following developmental exposure to Aroclor 1254 (Kodavanti et al., 2000). Developmental exposure to PCBs also caused significant hypothyroxinemia and age-dependent alterations in the translocation of PKC isozymes; the effects were significant at postnatal day (PND) 14 (Yang et al., 2003). Immunoblot analysis of PKC-alpha (α), -gamma (γ), and -epsilon (ε) from both cerebellum and hippocampus revealed that developmental exposure to Aroclor 1254 caused a significant decrease in cytosolic fraction and an increase in particulate fraction. For some isozymes, the ratio between the two fractions was increased in a dose-dependent manner. Thus, the patterns of subcellular distributions of PKC isoforms following a developmental PCB exposure were PKC isozyme- and developmental stage-specific. The changes in PKC and other second messengers were associated with changes in transcription factors such as Sp1 and NF-kB indicating changes in gene expression following developmental exposure to PCBs (Riyaz Basha et al., 2002). Considering the significant role of PKC signaling in motor behavior, learning and memory, it is suggested that altered subcellular distribution of PKC isoforms at critical periods of brain development may be associated with activation of transcription factors and subsequent gene expression, and may be a possible mechanism of PCB-induced neurotoxic effects. PKC-α, γ, and ε may be among the target molecules implicated with PCB-induced neurological impairments during developmental exposure.

Further studies focused on the structural outcome for changes in the intracellular signaling pathway following developmental PCB exposure. Detailed brain morphometric evaluation was performed by measuring neuronal branching and spine density. Figure 8.6 (upper panel) shows examples of hippocampal pyramidal and cerebellar Purkinje neurons. The lower panel in this figure shows the spines on the pyramidal neuron and a prototypic camera lucida drawing of the neuronal basilar tree.

Developmental exposure to PCBs affected normal dendritic development of Purkinje cells and CA1 pyramids (Mervis et al., 2002). The branching area was significantly smaller in the PCB-exposed rats (Table 8.4). When the rats became adults, there was continued neurostructural disruption of the CA1 dendritic arbor following PCB exposure, however, the branching area of the Purkinje cells returned

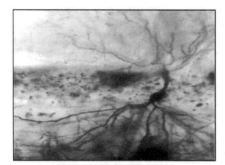

Hippocampal CA1 Pyramid

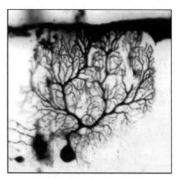

Cerebellum-Purkinje Cell

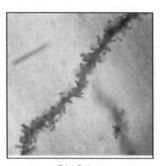

CA1 Spines

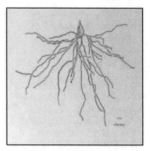

Camera Lucida Drawing
of CA1 Basilar Tree

Figure 8.6   Photomicrographs showing the morphology of neurons in hippocampus and cerebellum. Cerebral hemispheres were formalin-fixed for Rapid Golgi staining of tissue blocks which included the hippocampus and cerebellum, stained using the Golgi-Cox method. For branching and spine analysis of hippocampus, camera lucida drawings of the basilar dendritic tree were analyzed using the Sholl method of concentric circles. For dendritic spine analysis, counts were made along internal and terminal tip segments of neurons. For branching and spine analysis of cerebellar Purkinje cells, camera lucida drawings outlining the extent of the dendritic tree were made and the area of the dendritic domain was measured using a digitizing tablet. Dendritic branch density was assessed using an ocular grid. Dendritic spines were counted on tertiary terminal tip segments. (Courtesy of Dr. Ronald F. Mervis, Neurostructural Research Laboratory, Columbus, OH.) (Shown as Color Figure 8.6 following page 212.)

to normal level. Developmental exposure to PCBs also resulted in a significantly smaller spine density in hippocampus, but not in cerebellum (Table 8.4). This dysmorphic cytoarchitecture could be the structural basis for long-lasting neuro-cognitive deficits in PCB-exposed rats (Mervis et al., 2002).

Previously, Pruitt et al. (1999) reported a reduced growth of intra- and infra-pyramidal mossy fibers following developmental exposure to PCBs. These studies indicate that developmental exposure to a PCB mixture resulted in altered cellular distribution of PKC isoforms, which can subsequently disrupt the normal maintenance of signal transduction in developing neurons. The perturbations in intracellular signaling events could lead to structural changes in the brain. These findings suggest

Table 8.4 Effects of PCBs on Intracellular Signaling Processes and Brain Morphometry Following Developmental Exposure in Rats

Experimental Parameter	Cerebellum	Hippocampus
Calcium buffering[a,b]		
Microsomes	Altered	Altered
Mitochondria	Altered	Altered
Total PKC activity[a]	Inhibited	Not studied
PKC isoforms[c]		
PKC-alpha	Altered	Altered
PKC-gamma	Altered	Altered
PKC-epsilon	Altered	Altered
Transcription factors[d,e]		
Sp1	Increased	Increased
Ap-1	Increased	Increased
NF-kB	Increased	Increased
CREB	No change	Increased
Brain morphometry[f]		
Dendritic Branching	Decreased	Decreased
Spine density	No change	Decreased
Total PCBs[a]	2.19 ppm	Not determined

[a] Kodavanti et al., 2000.
[b] Sharma et al., 2000.
[c] Yang et al., 2003.
[d] Riyaz Basha et al., 2002.
[e] Unpublished data.
[f] Mervis et al., 2002.

that altered subcellular distribution of PKC isoforms may be a possible mode of action for PCB-induced neurotoxicity. Figure 8.5 illustrates the intracellular signaling events, transcription factors and brain morphometry affected by these chemicals.

Although the effects of PCBs on several intracellular second messengers and morphology at biologically and environmentally relevant concentrations have been demonstrated, it is very difficult to extrapolate these neurochemical changes to the neurotoxic effects of PCBs. In addition to the effects on second messengers, PCBs have been shown to decrease circulating thyroid hormones (Morse et al., 1993), decrease brain dopamine concentrations (Seegal et al., 1997), and alter the cholinergic system (Juarez de Ku et al., 1994). However, the mechanisms of their action, which lead to these changes, have not been investigated. Additional studies are needed to test whether one or more of these mechanisms are responsible for the effects of PCBs in the nervous system.

8.7 CONCLUSIONS

Developmental neurotoxicity involves alterations in behavior, neurophysiology, neurochemistry, and gross dysmorphology of the nervous system occurring in the offspring as a result of chemical exposure *in utero* and during lactation. Compared to adults, the developing organism has its own physiologic characteristics, uptake

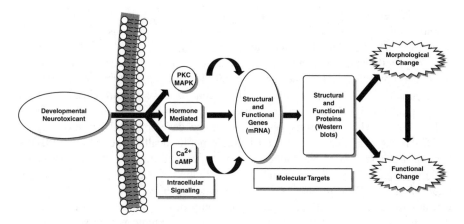

Figure 8.7 Schematic showing the proposed pathways (modes of action) by which a developmental neurotoxicant can cause a morphological or functional change.

characteristics, and possible inherent susceptibilities. The developing nervous system is particularly sensitive to toxicants, since there is rapid growth of the brain, known as "brain growth spurt." In rodents, this period is entirely neonatal (the first 3–4 weeks of life). In humans, the brain growth spurt begins during the third trimester of pregnancy and continues throughout the first years of life. During this period, the brain goes through several developmental processes starting from neurogenesis to proliferation, migration, and differentiation. Chemical exposure during any of these processes could have detrimental effects on the brain function and structure. Intracellular signaling pathways are crucial for the normal function and development of the nervous system. The intracellular second messengers generated in this process offer key steps in the mode of action for developmental neurotoxicants and provide good scientific basis for more quantitative and biologically defensible human health risk assessment. Figure 8.7 shows the proposed mode of action for persistent chemicals such as PCBs. Intracellular signaling bridges the gap between the chemical exposure and the adverse effect. The adverse effect could be a functional change or morphological change, although morphological changes may be a more sensitive indicator than a functional change. We now know that PCBs and structurally relevant chemicals alter intracellular signaling pathways such as calcium homeostasis, oxidative stress, PKC, and other kinase signaling systems. Alterations in these signaling pathways could lead to changes in the nuclear transcription factors which will be translated to altered structural and functional genes and their respective proteins. Alterations in these proteins could result in a morphologic or functional change (Figure 8.7). A number of developmental neurotoxicants could follow this mode of action.

Further, current methods used in hazard identification of developmental neurotoxicants rely on *in vivo* behavior and neuropathological measures to determine developmental neurotoxicity. It has been argued that these measures are highly variable and testing of neurochemical changes is limited (Claudio et al., 2000). The behavioral measures and gross neuropathology may not be sensitive enough to detect small changes caused by low environmental chemical exposure. This may be problematic, since humans are often exposed to low concentrations over a lifetime. Therefore, key events in the signal transduction pathways may lead to the development of specific and sensitive methods to detect developmental neurotoxicity based on mode(s) of action that can be used in hazard identification of untested chemicals. For example, PBDEs, that are structurally similar to PCBs (Figure 8.4), have been shown to affect signal transduction pathways similar to PCBs (Kodavanti and Derr-Yellin, 2002). Although human evidence for the developmental neurotoxicity of PBDEs is not available, there is considerable public health concern since PBDEs have been detected in human blood, adipose tissue, and breast milk. The levels are rapidly rising, and PBDEs are similar to PCBs in their effects on signal transduction (Kodavanti and Derr-Yellin, 2002). Understanding the effects of developmental neurotoxicants on intracellular signaling not only provides the mode of action for the neurotoxic effects but also provides biochemical and molecular markers that can be used in hazard identification of untested environmental chemicals.

ACKNOWLEDGMENTS

The authors are grateful for the graphic assistance of Chuck Gaul and for the editorial comments of Drs. Hugh Tilson and Stephanie Padilla of USEPA and Dr. Nasser Zawia of University of Rhode Island on an earlier version of this manuscript. This manuscript was reviewed by the NHEERL, U.S. Environmental Protection Agency, and approved for publication. Mention of trade names or commercial products does not constitute endorsement or recommendation for use. The opinions expressed by the authors are not to be misconstrued as U.S. Environmental Protection Agency policy.

REFERENCES

Abbott, N. J., Revest, P. A., and Romero, I.A. (1992), Astrocyte-endothelial interaction: physiology and pathology, *Neuropathol. Appl. Neurobiol.* 18, 424–433.

Abeliovich, A. et al. (1993), PKC-gamma mutant mice exhibit mild deficit in spatial and contextual learning, *Cell* 75, 1263–1271.

Adams, J. (1990), High incidence of intellectual deficits in 5 year old children exposed to isotretinoin "in utero," *Teratology* 41, 614.

Alexander, G. E., and Goldman, P. S. (1978), Functional development of the dorsolateral prefrontal cortex: an analysis utilizing reversible cryogenic depression, *Brain Res.* 143, 233–249.

Anderson, H. R., Nielsen, J. B., and Grandjean, P. (2000), Toxicologic evidence of developmental neurotoxicity of environmental chemicals, *Toxicology* 144, 121–127.

Andreoli, J., Rodier, P., and Langman, J. (1974), The influence of a prenatal trauma on formation of Purkinje cells, *Am. J. Anat.* 137, 87–102.

Angevine, J. B. Jr., and Sidman, R. L. (1961), Autoradiographic study of cell migration during histogenesis of cerebral cortex in the mouse, *Nature* 192, 766–768.

Angus, W. G., and Contreras, M. L. (1995), Aroclor 1254 alters the binding of ^{125}I-labeled nerve growth factor in PC12 cells, *Neurosci. Lett.* 191, 23–26.

Atkins, C. M. et al. (1998), The MAPK cascade is required for mammalian associative learning, *Nat. Neurosci.* 1, 602–609.

Audesirk, G. (1993), Electrophysiology of lead intoxication: effects on voltage-sensitive ion channels, *NeuroToxicology* 14, 137–148.

Balduini, W., Murphy, S. D., and Costa, L. G. (1987), Developmental changes in muscarinic receptor-stimulated phosphoinositide metabolism in rat brain, *J. Pharmacol. Exp. Ther.* 241, 421–427.

Balduini, W., and Costa, L. G. (1989), Effect of ethanol on muscarinic receptor stimulated phosphoinositide metabolism during brain development, *J. Pharmacol. Exp. Ther.* 250, 541–547.

Balduini, W., Candura, S. M., and Costa, L. G. (1991), Regional development of carbachol-, glutamate-, norepinephrine-, and serotonin-stimulated phosphoinositide metabolism in rat brain, *Dev. Brain Res.* 62, 115–120.

Bading, H. (1999), Nuclear calcium-activated gene expression: possible roles in neuronal plasticity and epileptogenesis, *Epilepsy Res.* 36, 225–231.

Bemis, J. C., and Seegal, R. F. (2000), Polychlorinated biphenyls and methylmercury alter intracellular calcium concentrations in rat cerebellar granule cells, *NeuroToxicology* 21, 1123–1134.

Berman, R. F., Verweij, B. H., and Muizelaar, J. P. (2000), Neurobehavioral protection by the neuronal calcium channel blocker, ziconotide, in a model of traumatic diffuse brain injury in rats, *J. Neurosurg.* 93, 821–828.

Besser, R. et al. (1987), Acute trimethyltin limbic cerebellar syndrome, *Neurology* 37, 945–950.

Bonventre, J. V. (1996), Roles of phospholipase A2 in brain cell and tissue injury associated with ischemia and excitotoxicity, *J. Lipid Med. Cell Signal.* 14, 15–23.

Burbacher, T. M., Rodier, P. M., and Weiss, B. (1990), Methylmercury developmental neurotoxicity: a comparison of effects in humans and animals, *Neurotoxicol. Teratol.* 12, 191–202.

Cabell, L., and Audesirk, G. (1993), Effects of selective inhibition of protein kinase C, cyclic AMP-dependent protein kinase, and Ca(2+)-calmodulin-dependent protein kinase on neurite development in cultured rat hippocampal neurons, *Int. J. Dev. Neurosci.* 11, 357–368.

Candura, S. M., Balduini, W, and Costa, L. G. (1991), Interaction of short chain aliphatic alcohols with muscarinic receptor-stimulated phosphoinositide metabolism in cerebral cortex from neonatal and adult rats, *NeuroToxicology* 12, 23–32.

Caputi, A. et al. (1996), Differential translocation of protein kinase C isozymes in rats characterized by a chronic lack of LTP induction and cognitive impairment, *FEBS Lett.* 393, 121–123.

Carafoli, E. (1987), Intracellular calcium homeostasis, *Annu. Rev. Biochem.* 56, 395–433.

Carpenter, D. O., Stoner, C. R., and Lawrence, D. A. (1997), Flow cytometric measurements of neuronal death triggered by PCBs, *NeuroToxicology* 18, 507–513.

Chalfie, M., and Wolinsky, E. (1990), The identification and suppression of inherited neurodegeneration in *Canorhabditis elegans*, *Nature* 345, 410–416.

Chen, S. J., Sweatt, J. D., and Klann, E. (1997), Enhanced phosphorylation of the postsynaptic protein kinase C substrate RC3/neurogranin during long-term potentiation, *Brain Res.* 749, 181–187.

Claudio, L. et al. (2000), Testing methods for developmental neurotoxicity of environmental chemicals, *Toxicol. Appl. Pharmacol.* 164, 1–14.

Claustre, Y., Rouguier, L., and Scatton, B. (1988), Pharmacological characterization of serotonin-stimulated phosphoinositide turnover in brain regions of the immature rat, *J. Pharmacol. Exp. Ther.* 244, 1051–1056.

Costa, L. G. (1993), Muscarinic receptors and the developing nervous system, *Receptors in the Developing Nervous System*, ed. I. S. Zagon and P. J. McLaughlin, Chapman and Hall, London, 21.

Costa, L. G. (1994), Second messenger systems in developmental neurotoxicity. *Developmental Neurotoxicology*, ed. G. J. Harry, CRC press, Boca Raton, FL, 77–101.

Costa, L. G. (1998a), Signal transduction in environmental neurotoxicity, *Annu. Rev. Pharmacol. Toxicol.* 38, 21–43.

Costa, L. G. (1998b), Ontogeny of second messenger systems, *Handbook of Developmental Neurotoxicology*, ed. W. L. Slikker and L. W. Chang, Academic Press, San Diego, CA, 275–284.

Costa, L. G. et al. (2001), Intracellular signal transduction pathways as targets for neurotoxicants, *Toxicology* 160, 19–26.

Davis, J. M. et al. (1990), The comparative developmental neurotoxicity of lead in humans and animals, *Neurotoxicol. Teratol.* 12, 215–229.

Di Luca, M., Caputi, A., and Cattabeni, F. (1994), Synaptic protein phosphorylation changes in animals exposed to neurotoxicants during development, *NeuroToxicology* 15, 525–532.

Dorner, G, and Plagemann, A. (2002), DDT in human milk and mental capacities in children at school age: an additional view on PISA 2000, *Neuroendocrinol. Lett.* 23, 427–431.

Driscoll, C. D., Streissguth, A. P., and Riley, E. P. (1990), Prenatal alcohol exposure: comparability of effects in humans and animal models, *Neurotoxicol. Teratol.* 12, 231–237.

Edwards, D. R. (1994), Cell signaling and the control of gene transcription, *Trends Pharmacol. Sci.* 15, 239–244.

English, D. (1996), Phosphatidic acid: a lipid messenger involved in intracellular and extracellular signaling, *Cell Signal.* 8, 341–347.

Eriksson, P. (1997), Developmental neurotoxicity of environmental agents in the neonate, *NeuroToxicology* 18, 719–726.

Erickson, M. D. (1986), *Analytical Chemistry of PCBs*, Butterworth, Boston.

Evans, P. H. (1993), Free radicals in brain metabolism and pathology, *Br. Med. Bull.* 49, 577–587.

Exton, J. H. (1996), Regulation of phosphoinositide phospholipases by hormones, neurotransmitters and agonist linked to G proteins, *Annu. Rev. Pharmacol. Toxicol.* 36, 481–509.

Faustman, E. M. et al. (2002), Investigations of methylmercury-induced alterations in neurogenesis, *Environ. Health Perspect.* 110, 859–864.

Ferrell, J. E. (1996), MAP kinases in mitogenesis and development, *Curr. Top. Dev. Biol.* 33, 1–60.

Fischer, L. J. et al. (1998), Symposium overview: Toxicity of non-coplanar PCBs, *Toxicol. Sci.* 41, 49–61.

Fisher, S. K., Heacock, A. M., and Agranoff, B. W. (1992), Inositol lipids and signal transduction in the nervous system: an update, *J. Neurochem.* 58, 18–38.

Fitzjohn, S. M., and Collingridge, G. L. (2002), Calcium stores and synaptic plasticity, *Cell Calcium* 32, 405–411.

Gibson, G. E., and Peterson, C. (1987), Calcium and the aging nervous system, *Neurobiol. Aging* 8, 329–343.

Gilbert, M. E., and Crofton, K. M. (1999), Developmental exposure to a commercial PCB mixture (Aroclor 1254), produces a persistent impairment in long-term potentiation in rat dentate gyrus *in vivo, Brain Res.* 850, 87–95.

Girard, P. R., and Kuo, J. F. (1990), Protein kinase C and its 80-kilodalton substrate protein in neuroblastoma cell neurite outgrowth, *J. Neurochem.* 54, 300–306.

Godinho, A. F., Trombini, T. V., and Oliveira, E. C. (2002), Effects of elevated calcium on motor and exploratory activities of rats, *Braz. J. Med. Biol. Res.* 35, 451–457.

Goldey, E. S. et al. (1994), Developmental neurotoxicity: evaluation of testing procedures with methylazoxymethanol and methylmercury, *Fundam. Appl. Toxicol.* 23, 447–464.

Goldey, E. S., Tilson, H. A., and Crofton, K. M. (1995), Implications of the use of neonatal birth weight, growth, viability, and survival data for predicting developmental neurotoxicity: a survey of the literature, *Neurotoxicol. Teratol.* 17, 313–332.

Goldstein, G. W. (1993), Evidence that lead acts as a calcium substitute in second messenger metabolism, *NeuroToxicology* 14, 97–102.

Gore, A. C. et al. (2002), A novel mechanism for endocrine-disrupting effects of polychlorinated biphenyls: direct effects on gonadotropin-releasing hormone neurons, *J. Neuroendocrinol.* 14, 814–823.

Grant, K. A., and Samson, H. H. (1982), Ethanol and tertiary butanol induced microencephaly in the neonatal rat: comparison of brain growth parameters, *Neurobehav. Toxicol. Teratol.* 4, 315–321.

Grant, K. A., and Samson, H. H. (1984), N-Propanol induced microencephaly in the neonatal rat, *Neurobehav. Toxicol. Teratol.* 6, 165–169.

Hall, A. (1990), Oncogene products involved in signal transduction, *G-proteins and Calcium Signaling*, ed. P. H. Naccache, CRC Press, Boca Raton, FL, 29.

Hashimoto, T. et al. (1988), Postnatal development of a brain-specific subspecies of protein kinase C in rat, *J. Neurosci.* 8, 1678–1683.

Haykal-Coates, N. et al. (1998), Effects of gestational methylmercury exposure on immunoreactivity of specific isoforms of PKC and enzyme activity during postnatal development of the rat brain, *Dev. Brain Res.* 109, 33–49.

Heiss, W. D. (1981), Cerebral blood flow: physiology, pathophysiology and pharmacological effects, *Adv. Otorhinolaryngol.* 27, 26–39.

Hoek, J. B., and Rubin, E. (1990), Alcohol and membrane-associated signal transduction, *Alcohol Alcohol* 25, 143–156.

Hoffman, P. L., and Tabakoff, B. (1990), Ethanol and guanine nucleotide binding proteins: selective interaction, *FASEB J.* 4, 2612–2622.

Hoffman, P. L., and Tabakoff, B. (1996), To be or not to be: how ethanol can affect neuronal death during development, *Alcohol Clin. Exp. Res.* 20, 193–195.

Honegger, P. (1986), Protein kinase C-activating tumor promoters enhance the differentiation of astrocytes in aggregating fetal brain cell cultures, *J. Neurochem.* 46, 1561–1566.

Howard, E. (1973), DNA content of rodent brains during maturation and aging and autoradiography of postnatal DNA synthesis in monkey brain, *Prog. Brain Res.* 40, 91–114.

Inglefield, J. R., and Shafer, T. J. (2000), Polychlorinated biphenyl-stimulation of Ca^{2+} oscillations in developing neocortical cells: a role for excitatory transmitters and L-type voltage-sensitive Ca^{2+} channels, *J. Pharmacol. Exp. Ther.* 295, 105–113.

Inglefield, J. R. et al. (2002), Identification of calcium-dependent and -independent signaling pathways involved in polychlorinated biphenyl-induced cyclic AMP-responsive element-binding protein phosphorylation in developing cortical neurons, *Neuroscience* 115, 559–573.

Jacobson, J. L., and Jacobson, S. W. (1996), Intellectual impairment in children exposed to polychlorinated biphenyls *in utero*, *New Engl. J. Med.* 335, 783–789.

Jerusalinsky, D. et al. (1994), Post-training intrahippocampal infusion of protein kinase C inhibitors causes amnesia in rats, *Behav. Neural. Biol.* 61, 107–109.

Juarez de Ku, L. M., Sharma-Stokkermans, M., and Meserve, L. A. (1994), Thyroxine normalizes polychlorinated biphenyl (PCB), dose-related depression of choline acetyltransferase (ChAT), activity in hippocampus and basal forebrain of 15-day old rats, *Toxicology* 94, 19–30.

Kafafi, S. A. et al. (1993), Binding of polychlorinated biphenyls to the aryl hydrocarbon receptor, *Environ. Health Perspect.* 101, 422–428.

Kakita, A. et al. (2002), Neuronal migration disturbance and consequent cytoarchitecture in the cerebral cortex following transplacental administration of methylmercury, *Acta Neuropathol.* 104, 409–417.

Kanthasamy, A. G. et al. (1994), Dopaminergic neurotoxicity of cyanide: neurochemical, histological, and behavioral characterization, *Toxicol. Appl. Pharmacol.* 126, 156–163.

Kater, S. B., and Mills, L. R. (1991), Regulation of growth cone behavior by calcium, *J. Neurosci.* 11, 891–899.

Katz, B. (1969), *The Release of Neural Transmitter Substances*, Liverpool University Press.

Kaul, P. P. et al. (1996), Fenvalerate-induced alterations in circulatory thyroid hormones and calcium stores in rat brain, *Toxicol. Lett.* 89, 29–33.

Kater, S. B., and Mills, L. R. (1991), Regulation of growth cone behavior by calcium. *J. Neurosci.* 11, 891–899.

Kern, M., and Audesirk, G. (1995), Inorganic lead may inhibit neurite development in cultured rat hippocampal neurons through hyperphosphorylation, *Toxicol. Appl. Pharmacol.* 134, 111–123.

Keshles, O., and Levitzki, A. (1984), The ontogeny of beta-adrenergic receptors and of adenylate cyclase in the developing rat brain, *Biochem. Pharmacol.* 33, 3231–3233.

Khan, W. A., Blobe, G. C., and Hannun, Y. A. (1995), Arachidonic acid and free fatty acids as second messengers and the role of protein kinase C, *Cell Signal.* 3, 171–184.

Kodavanti, P. R. S. et al. (1993), Comparative effects of two polychlorinated biphenyl congeners on calcium homeostasis in rat cerebellar granule cells, *Toxicol. Appl. Pharmacol.* 123, 97–106.

Kodavanti, P. R. S. et al. (1994), Differential effects of polychlorinated biphenyl congeners on phosphoinositide hydrolysis and protein kinase C translocation in rat cerebellar granule cells, *Brain Res.* 662, 75–82.

Kodavanti, P. R. S. et al. (1996a), Inhibition of microsomal and mitochondrial Ca^{2+} sequestration in rat cerebellum by polychlorinated biphenyl mixtures and congeners: structure-activity relationships, *Arch. Toxicol.* 70, 150–157.

Kodavanti, P. R. S. et al. (1996b), Increased [^{3}H]phorbol ester binding in rat cerebellar granule cells and inhibition of $^{45}Ca^{2+}$ sequestration by polychlorinated diphenyl ether congeners and analogs: structure-activity relationships, *Toxicol. Appl. Pharmacol.* 138, 251–261.

Kodavanti, P. R. S, and Tilson, H. A. (1997), Structure-activity relationships of potentially neurotoxic PCB congeners in the rat, *NeuroToxicology* 18, 425–442.

Kodavanti, P. R. S., and Derr-Yellin, E. C. (1999), Activation of calcium-dependent and -independent phospholipase A_2 by non-coplanar polychlorinated biphenyls in rat cerebellar granule neurons, *Organohalogen Comp.* 42, 449–453.

Kodavanti, P. R. S., and Tilson, H. A. (2000), Neurochemical effects of environmental chemicals: *in vitro* and *in vivo* correlations on second messenger pathways, *Ann. N. Y. Acad. Sci.* 919, 97–105.

Kodavanti, P. R. S. et al. (2000), Developmental exposure to Aroclor 1254 alters calcium buffering and protein kinase C activity in the brain, *Toxicol. Sci.* 54, 76–77.

Kodavanti, P. R. S., and Derr-Yellin, E. C. (2002), Differential effects of polybrominated diphenyl ethers and polychlorinated biphenyls on [^{3}H]arachidonic acid release in rat cerebellar granule neurons, *Toxicol. Sci.* 68, 451–457.

Krebs, E. G., (1989), Role of cyclic AMP-dependent protein kinase in signal transduction, *JAMA* 262, 1815–1818.

Lanier, L. P., Dunn, A. J., and Van Hartesveldt, C. (1976), Development of neurotransmitters and their formation in brain, *Rev. Neurosci.* 2, 195–256.

Levy, M. et al. (2003), Mitochondrial regulation of synaptic plasticity in the hippocampus, *J. Biol. Chem.* 278, 17727–17734.

Limke, T. L., Otero-Montanez, J. K., and Atchison, W. D. (2003), Evidence for interactions between intracellular calcium stores during methylmercury-induced intracellular calcium dysregulation in rat cerebellar granule neurons, *J. Pharmacol. Exp. Ther.* 304, 949–958.

Liscovitch, M. (1996), Phospholipase D: role in signal transduction and membrane traffic, *J. Lipid Med. Cell. Signal.* 14, 215–221.

Liu J-P. (1996), Protein kinase C and its substrates, *Mol. Cell Endocrinol.* 116, 1–29.

Llinas, R. R. (1988), The electrophysiological properties of mammalian neurons: insights into central nervous system function, *Science* 242, 1654–1664.

Long, G. J., Rosen, J. F., and Schanne, F. A. X. (1994), Lead activation of protein kinase C from rat brain, *J. Biol. Chem.* 269, 834–837.

LoPachin, R. M., and Aschner, M. (1999), Neuronal-glial interactions as potential targets of neurotoxicant effect, *Neurotoxicology*, 2nd ed., ed. H. A. Tilson and G. J. Harry, Taylor & Francis, Philadelphia, PA, 53–80.

Lynch, M. A. (1998), Analysis of the mechanisms underlying the age-related impairment in long-term potentiation in the rat, *Rev. Neurosci.* 9, 169–201.

Ma, F. H. et al. (1991), Ontogeny of beta-adrenergic receptor-mediated cyclic AMP generating system in primary cultured neurons, *Int. J. Dev. Neurosci.* 9, 347–356.

Maduh, E. U. et al. (1990), Cyanide-induced neurotoxicity: calcium mediation of morphological changes in neuronal cells, *Toxicol. Appl. Pharmacol.* 103, 214–221.

Miale, I. L., and Sidman, R.L. (1961), An autoradiographic analysis of histogenesis in the mouse cerebellum, *Exp. Neurol.* 4, 277–296.

Majewski, H., and Iannazzo, L. (1998), Protein kinase C: a physiological mediator of enhanced transmitter output, *Prog. Neurobiol.* 55, 463–475.

Malarkey, K. et al. (1995), The regulation of tyrosine kinase signaling pathways by growth factor and G-protein-coupled receptors, *Biochem. J.* 309, 361–375.

Malenka, R. C. et al. (1989), The impact of postsynaptic calcium on synaptic transmission — its role in long-term potentiation, *Trends Neurosci.* 12, 444–450.

Marty, A. (1989), The physiological role of calcium dependent channels, *Trends Neurosci.* 12, 420–424.

Matsushima, H. et al. (1996), Ca^{2+}-dependent and Ca^{2+}-independent protein kinase C changes in the brain of patients with Alzheimer's disease, *J. Neurochem.* 67, 317–323.

Mattson, M. P. (1988), Neurotransmitters in the regulation of neuronal cytoarchitecture, *Brain Res.* 472, 179–212.

Mattson, M. P., Guthrie, P. B., and Kater, S. B. (1988), Intracellular messengers in the generation and degeneration of hippocampal neuroarchitecture, *J. Neurosci. Res.* 21, 447–464.

Mayer, P. et al. (2002), Gene expression profile after intense second messenger activation in cortical primary neurons, *J. Neurochem.* 82, 1077–1086.

McCann, S. M. et al. (1998), Hypothalamic control of FSH and LH by FSH-RF, LHRH, cytokines, leptin and nitric oxide, *Neuroimmunomodulation* 5, 193–202.

McCormack, J. G., Halestrap, A. P., and Denton, R. M. (1990), Role of calcium ions in the regulation of mammalian intramitochondrial metabolism, *Physiol. Rev.* 70, 391–425.

McDonald, J. W., and Johnston, M. V. (1990), Physiological and pathophysiological roles of excitatory amino acids during central nervous system development, *Brain Res. Rev.* 15, 41–70.

McKinney, J. D. and Waller, C. L. (1994), Polychlorinated biphenyls as hormonally active structural analogues, *Environ. Health Perspect.* 102, 290–297.

Mervis, R. F. et al. (2002), Long-lasting neurostructural consequences in the rat hippocampus by developmental exposure to a mixture of polychlorinated biphenyls, *Toxicol. Sci.* 66, 133.

Miller R. J. (1991), The control of neuronal Ca^{2+} homeostasis, *Prog. Neurobiol.* 37, 255–285.

Morgan, J. F., and Curran, T. (1989), Stimulus-transcription coupling in neurons: role of immediate early genes, *Trends Neurosci.* 12, 459–462.

Morse, D. C. et al. (1993), Interference of PCBs in hepatic and brain thyroid hormone metabolism in fetal and neonatal rats, *Toxicol. Appl. Pharmacol.* 122, 27–33.

Mosquera, D. I., Stedeford, T., Cardozo-Pelaez, F., and Sanchez-Ramos, J. (2003), Strain-specific differences in the expression and activity of Ogg1 in the CNS, *Journal of Gene Expression* 11(1), 47–53.

Mundy, W. R., Freudenrich, T. M., and Kodavanti, P. R. S. (1997), Aluminum potentiates glutamate-induced calcium accumulation and iron-induced oxygen free radical formation in primary neuronal cultures, *Mol. Chem. Neuropathol.* 32, 41–57.

Mundy, W. R. et al. (1999), Extracellular calcium is required for the polychlorinated biphenyl-induced increase of intracellular free calcium levels in cerebellar granule cell culture, *Toxicology* 136, 27–39.

Mundy, W. R., and Freudenrich, T. M. (2000), Sensitivity of immature neurons in culture to metal-induced changes in reactive oxygen species and intracellular free calcium, *NeuroToxicology* 21, 1135–1144.

Murphy, S. et al. (1987), Phorbol ester stimulates proliferation of astrocytes in primary cultures, *Brain Res.* 428, 133–135.

Nakai, A. et al. (2000), Developmental changes in mitochondrial activity and energy metabolism in fetal and neonatal rat brain, *Dev. Brain Res.* 121, 67–72.

Narahashi, T. (2000), Neuroreceptors and ion channels as the basis for drug action: past, present, and future, *J. Pharmacol. Exp. Ther.* 294, 1–26.

National Research Council. (2000), *Scientific Frontiers in Developmental Toxicology and Risk Assessment*, National Academy Press, Washington, DC.

Nichlas, W. J. et al. (1992), Mitochondrial mechanisms of neurotoxicity, *Ann. N. Y. Acad. Sci.* 648, 28–36.

Niemi, W. D. et al. (1998), PCBs reduce long-term potentiation in the CA1 region of rat hippocampus, *Exp. Neurol.* 151, 26–34.

Nihei, M. K. et al. (2001), Low level Pb^{2+} exposure affects hippocampal protein kinase-gamma gene and protein expression in rats, *Neurosci. Lett.* 298, 212–216.

Nishizuka, Y. (1995), Protein kinase C and lipid signaling for sustained cellular responses, *FASEB J.* 9, 489–496.

Okey, A. B., Riddick, D. S., and Harper, P. A. (1994), The Ah receptor: mediator of the toxicity of 2,3,7,8-tetrachlorodibezo-p-dioxin (TCDD), and related compounds, *Toxicol. Lett.* 70, 1–22.

Oyama, Y. et al. (1994), Methylmercury-induced augmentation of oxidative metabolism in cerebellar neurons dissociated from the rats: its dependence on intracellular Ca^{2+}, *Brain Res.* 660, 154–157.

Palmer, E. et al. (1990), Changes in excitatory amino acid modulation of phosphoinositide metabolism during development, *Dev. Brain Res.* 51, 132–134.

Parran, D. K., Barone, S., Jr., and Mundy, W. R. (2003), Methylmercury decreases NGF-induced TrkA autophosphorylation and neurite outgrowth in PC12 cells, *Dev. Brain Res.* 141, 71–81.

Patandin, S. et al. (1999), Effects of environmental exposure to polychlorinated biphenyls and dioxins on cognitive abilities in Dutch children at 42 months of age, *J. Pediat.* 134, 33–41.

Pavlakovic, G. et al. (1995), Activation of protein kinase C by trimethyltin: relevance to neurotoxicity, *J. Neurochem.* 65, 2338–2343.

Pendlebury, W. W. et al. (1987), Results of immunocytochemical, neurochemical and behavioral studies in aluminum-induced neurofilamentous degeneration, *J. Neural Trans.* 24, 213–217.

Pierce, E. T. (1973), Time of origin of neurons in the brain stem of the mouse, *Prog. Brain Res.* 40, 53–65.

Pounds, J. G. (1984), Effects of lead intoxication on calcium homeostasis and calcium-mediated cell function: a review, *NeuroToxicology* 5, 295–332.

Pruitt, D. L., Meserve, L. A., and Bingman, V. P. (1999), Reduced growth of intra- and infra-pyramidal mossy fibers is produced by continuous exposure to polychlorinated biphenyl, *Toxicology* 138, 11–17.

Rajanna, B. et al. (1995), Modulation of protein kinase C by heavy metals, *Toxicol. Lett.* 81, 197–203.

Rakic, P., and Sidman, R. L. (1970), Histogenesis of cortical layers in human cerebellum, particularly the lamina dissecans, *J. Comp. Neurol.* 139, 473–500.

Reboulleau, C. P. (1986), Extracellular calcium-induced neuroblastoma cell differentiation: involvement of phosphatidylinositol turnover, *J. Neurochem.* 46, 920–930.

Reinholz, M. M., Bertics, P. J., and Miletic, V. (1999), Chronic exposure to lead acetate affects the development of protein kinase C activity and the distribution of the PKCgamma isozyme in the rat hippocampus, *NeuroToxicology* 20, 609–617.

Risau, W., and Wolburg, H. (1990), Development of the blood-brain barrier, *Trends Neurosci.* 13, 714–778.

Riyaz Basha, M. et al. (2002), Changes in the DNA-binding of several transcription factors in the developing rat cerebellum by PCBs, *Organohal. Comp.* 57, 381–383.

Rodier, P. M. (1980), Chronology of neuron development: animal studies and their clinical implications, *Dev. Med. Child Neurol.* 22, 525–545.

Safe, S. (1994), Polychlorinated biphenyls (PCBs): environmental impact, biochemical and toxic responses, and implications for risk assessment, CRC *Crit. Rev. Toxicol.* 24, 87–149.

Saunders, R. D., and De Vries, G. H. (1988), Schwann cell proliferation is accompanied by enhanced inositol phosphate metabolism, *J. Neurochem.* 50, 876–882.

Schantz, S. L. (1996), Developmental neurotoxicity of PCBs in humans: what do we know and where do we go from here, *Neurotoxicol. Teratol.* 18, 217–227.

Schantz, S. L., Moshtaghian, J., and Ness, D. K. (1995), Spatial learning deficits in adult rats exposed to *ortho*-substituted PCB congeners during gestation and lactation, *Fundam. Appl. Toxicol.* 26, 117–126.

Schmidt, M. J., Palmer, G. C., and Robinson, G. A. (1980), The cyclic nucleotide system in the brain during development and aging, *Psychopharmacology of Aging*, ed. C. Eisdorfer and W. E. Fann, Spectrum, New York, 213–265.

Schoeppe, D. D., and Rutledge, C. O. (1985), Comparison of postnatal changes in alpha-adrenoceptor binding and adrenergic stimulation of phosphoinositide hydrolysis in rat cerebral cortex, *Biochem. Pharmacol.* 34, 2705–2711.

Schull, W. J., Norton, S., and Jensh, R. P. (1990), Ionizing radiation and the developing brain, *Neurotoxicol. Teratol.* 12, 249–260.

Schuman, E. M. and Madison, D. V. (1994), Nitric oxide and synaptic function, *Annu. Rev. Neurosci.* 17, 153–183.

Seegal, R. F., Brosch, K. O., and Okoniewski, R. J. (1997), Effects of *in utero* and lactational exposure of the laboratory rat to 2,4,2′,4′- and 3,4,3′,4′-tetrachlorobiphenyl on dopamine function, *Toxicol. Appl. Pharmacol.* 146, 95–103.

Serrano, P. A. et al. (1994), Differential effects of protein kinase inhibitors and activators on memory formation in the 2-day-old chick, *Behav. Neural Biol.* 61, 60–72.

Shafer, T. J. et al. (1996), Disruption of inositol phosphate accumulation in cerebellar granule cells by polychlorinated biphenyls: a consequence of altered Ca^{2+} homeostasis, *Toxicol. Appl. Pharmacol.* 141, 448–455.

Shapiro, D. L. (1973), Morphological and biochemical alterations in fetal rat brain cells cultured in the presence of monobutyryl cyclic AMP, *Nature* (London), 241, 203–204.

Sharma, R., and Kodavanti, P. R. S. (2002), *In vitro* effects of polychlorinated biphenyls and hydroxy metabolites on nitric oxide synthases in rat brain, *Toxicol. Appl. Pharmacol.* 178, 127–136.

Sharma, R., Derr-Yellin, E. C., House, D. E., and Kodavanti, P. R. S. (2000). Age-dependent effects of Aroclor 1254 on calcium uptake by subcellular organelles in selected brain regions of rats. *Toxicology* 156, 13–25.

Shearman, M. S. et al. (1991), Synaptosomal protein kinase C subspecies. A. Dynamic changes in the hippocampus and cerebellar cortex concomitant with synaptogenesis, *J. Neurochem.* 56, 1255–1262.

Sheng, M., and Greenberg, M. E. (1990), The regulation and formation of c-fos and other immediate early genes in the nervous system, *Neuron* 4, 477–485.

Sidman, R. I. (1969), Autoradiographic methods and principles for study of the nervous system with thymidine-3H, *Contemporary Research Methods in Neuroanatomy*, ed. W. J. H. Nauta and S. O. E. Ebbesson, Springer-Verlag, New York, 252–274.

Siesjo, B. K. (1990), Calcium in the brain under physiological and pathological conditions, *Eur. Neurol.* 30, 3–9.

Simon, M. I., Strathmann, M. P., and Gautam, N. (1991), Diversity of G proteins in signal transduction, *Science* 252, 802–808.

Sirois, J. E., and Atchison, W. D. (2000), Methylmercury affects multiple subtypes of calcium channels in rat cerebellar granule cells, *Toxicol. Appl. Pharmacol.* 167, 1–11.

Smith, S. J., and Augustine, G. J. (1988), Calcium ions, active zones and synaptic transmitter release, *Trends Neurosci.* 11, 458–464.

Spohr, H. L., Williams, J., and Steinhausen, H. C. (1993), Prenatal alcohol exposure and long-term developmental consequences, *Lancet* 341, 907–909.

Sposi, N. M. et al. (1989), Expression of protein kinase C genes during ontogenetic development in the central nervous system, *Mol. Cell Biol.* 9, 2284–2288.

Steketee, J. D. (1993), Injection of the protein kinase inhibitor H7 into the A10 dopamine region blocks the acute responses to cocaine: behavioral and *in vivo* microdialysis studies, *Neuropharmacology* 32, 1289–1297.

Streissguth, A. P. et al. (1980), Teratogenic effects of alcohol in humans and laboratory animals, *Science* 209, 353–361.

Striggow, F., and Ehrlich, B. E. (1997), Regulation of intracellular calcium release channel function by arachidonic acid and leukotriene B4, *Biochem. Biophys. Res. Commun.* 237, 413–418.

Sukhotina, I. A. et al. (1999), Effects of calcium channel blockers on behaviors induced by the N-methyl-D-aspartate receptor antagonist, dizocilpine, in rats, *Pharmacol. Biochem. Behav.* 63, 569–580.

Sunayashi-Kusuzaki, K. et al. (1993), Associative learning potentiates protein kinase C activation in synaptosomes of the rabbit hippocampus, *Proc. Natl. Acad. Sci. USA* 90, 4286–4289.

Szekely, A. M., Costa, E. and Gayson, D. R. (1990), Transcriptional program coordination by N-methyl-D-aspartate sensitive glutamate receptor stimulation in primary cultures of cerebellar neurons, *Mol. Pharmacol.* 38, 624–633.

Tanaka, C., and Nishizuka, Y. (1994), The protein kinase C family for neuronal signaling, *Annu. Rev. Neurosci.* 17, 551–567.

Theiler, K. (1972), *The House Mouse: Development and Normal Stages from Fertilization to 4 Weeks of Age*, Springer-Verlag, New York.

Tilson, H. A., Jacobson, J. L., and Rogan, W. J. (1990), Polychlorinated biphenyls and the developing nervous system: cross-species comparisons, *Neurotoxicol. Teratol.* 12, 239–248.

Tilson, H. A., and Kodavanti, P. R. S. (1997), Neurochemical effects of polychlorinated biphenyls: an overview and identification of research needs, *NeuroToxicology* 18, 727–744.

Tilson, H. A., and Kodavanti, P. R. S. (1998), The neurotoxicity of polychlorinated biphenyls, *NeuroToxicology* 19, 517–526.

Traynor, A. E. (1984), The relationship between neurite extension and phospholipid metabolism in PC12 cells, *Brain Res.* 316, 205–210.

Trilivas, I., and Brown, J. H. (1989), Increases in intracellular Ca^{2+} regulate the binding of [^{3}H]phorbol 12,13-dibutyrate to intact 1321N1 astrocytoma cells, *J. Biol. Chem.* 264, 3102–3107.

Voie, O. A., and Fonnum, F. (1998), *Ortho*-substituted polychlorinated biphenyls elevate intracellular [Ca^{2+}] in human granulocytes, *Environ. Toxicol. Pharmacol.* 5, 105–112.

Vorhees, C. V. (1992), Developmental neurotoxicology, *Neurotoxicology*, ed. H. A. Tilson and C. Mitchell, Raven Press, New York, 295–329.

Walsh, T. J., and DeHaven, D. L. (1988), Neurotoxicity of the alkyltins, *Metal Neurotoxicity*, ed. S. C. Bondy and K. N. Presad, CRC Press, Boca Raton, FL, 87–108.

Wang, X., and Robinson, P. J. (1997), Cyclic-GMP-dependent protein kinase and cellular signaling in the nervous system, *J. Neurochem.* 68, 443–456.

White, D. M. et al. (1992), Neurologic syndrome in 25 workers from an aluminum smelting plant, *Arch. Intern. Med.* 152, 1443–1448.

Woodward, D. J., Hoffer, B. J., and Lapham, L. W. (1969), Correlative survey of electrophysiological, neuropharmacological, and histochemical aspects of cerebellar maturation in the rat, *Neurobiology of Cerebellar Evolution and Development*, ed. R. Llinas, American Medical Association, Chicago, 725–741.

World Health Organization (1993), *Environmental Health Criteria 140: Polychlorinated Biphenyls and Terphenyls*, 2nd ed., prepared by S. Dobson and G. J. van Esch, International Programme on Chemical Safety, Geneva, Switzerland.

Yamamoto, H., (1990), Protection against cyanide-induced convulsions with alpha-ketoglutarate, *Toxicology* 61, 221–228.

Yamashita, F., and Hayashi, M. (1985), Fetal PCB syndrome: clinical features, intrauterine growth retardation and possible alteration in calcium metabolism, *Environ. Health Perspect.* 59, 41–45.

Yang, J-H., and Kodavanti, P. R. S. (2001), Possible molecular targets of halogenated aromatic hydrocarbons in neuronal cells, *Biochem. Biophys. Res. Commun.* 280, 1372–1377.

Yang, J-H., Derr-Yellin, E. C., and Kodavanti, P. R. S. (2003), Alterations in brain protein kinase C isoforms following developmental exposure to a polychlorinated biphenyl mixture, *Mol. Brain Res.* 111, 123–135.

Yong, V. W. et al. (1988), Phorbol ester enhances morphological differentiation of oligodendrocytes in culture, *J. Neurosci. Res.* 19, 187–194.

Zhang, J., and Snyder, S. H. (1995), Nitric oxide in the nervous system, *Annu. Rev. Pharmacol. Toxicol.* 35, 213–233.

CHAPTER 9

The Role of POU Domain Transcription Factors in Lead Neurotoxicity

Saleh A. Bakheet and Nasser H. Zawia

CONTENTS

9.1 INTRODUCTION

Lead exposure and toxicity continue to be a major public health problem in urban centers in the U.S. and around the world. Human exposure to lead, especially at a young age, adversely affects a range of body systems including the reproductive, gastrointestinal, immune, renal, cardiovascular, skeletal, muscular, hematopoetic, and nervous systems (Goyer, 1996). In particular, the central nervous system (CNS) is sensitive to lead (Finkelstein et al., 1998). In October 1991 the Centers for Disease Control (CDC) published a new statement on the prevention of lead poisoning in young children which revised the 1985 intervention blood lead level of 25 µg/dl downward to 10 µg/dl (0.483 µM/l) (CDC, 1992). Approximately 8.9% of children in the U.S. alone exceed that threshold (Goldstein, 1992; Needleman, 1994, 1998; Bressler et al., 1999). In children, blood lead level as low as 0.5–1.0 µM may affect CNS development leading to mental retardation and permanent cognitive deficits (Bellinger et al., 1987). Any exposure to lead is harmful to the CNS and a threshold below which lead has no effect has yet to be discovered (Lidsky and Schneider, 2003).

Children are more susceptible to the effects of lead exposure due to several factors, including enhanced overall absorption and decreased excretion of lead as well as greater brain uptake because of the immaturity of the blood–brain barrier (BBB). Additionally, the behavior of children (e.g., playing outdoors, hand-to-mouth activity) increases their risk of exposure (Duggan and Inskip, 1985). Studies in human pediatric populations have established that exposures producing blood lead level as low as 10 µg/dl can deleteriously impact cognitive functions as measured by changes in intelligence quotient (IQ) test scores and other psychometric indices (WHO, 1994). Early case reports described permanent mental retardation and impaired cognitive functions as a result of acute high level lead exposure in children (Byers and Lord, 1943; Perlstein and Attala, 1966). Cognitive deficits in lead-exposed children have been documented at much lower lead levels. A recent report confirms that deficits in cognitive and academic skills are associated with lead exposure at blood lead concentrations lower than 5 µg/dl (Lanphear et al., 2000). There is a direct link between low level exposure and deficits in the neurobehavioral cognitive performance in childhood through adolescence. However, exposure to lead is a preventable childhood disease (Canfield et al., 2003).

The world population continues to sustain lifetime exposure to lead due to residual contamination of dust, soil, food, and water supply from the many years of use of lead-based paints and gasoline. Efforts to reduce environmental lead exposure include the removal of lead from paint and gasoline, the regulation of lead in drinking water, the prohibition of lead solder in food cans, and safe and effective lead removal programs in houses (Pirkle et al., 1994). These efforts have resulted in a significant reduction in blood lead concentrations. Although lead consumption is lower today, approximately 400,000 metric tons of lead are still used each year in the U.S., and lead appears in a wide array of consumer products including batteries, solder, galvanized pipes, ammunition, roofing materials, and x-ray shielding (Florini et al., 1990; ATSDR, 1993). According to Pirkle (1994), in 1989 approximately 22% of the U.S. population occupied houses that were built before 1940, when lead-based paint was commonly used. Despite a large decline in the number of housing units with lead-based paint from 1990 to 2000, there are still millions of houses remaining with this hazard (Jacobs et al., 2002).

The health hazards resulting from lead exposure are numerous; however, the impact of lead on normal brain development is of utmost importance. One way to try to treat or reverse the adverse effects of lead on brain development is to understand the molecular events that are damaged by lead. There are multiple cellular targets in the developing brain at which lead may act. Lead is able to interfere with several signal transduction pathways and transcription factors, thereby resulting in alterations in gene expression. A potential mechanism through which environmental agents could precipitate cerebral damage is by perturbations in the regulation of genes mediated by transcription factors involved in neural development. In this chapter, we will discuss the toxic effects of lead in the developing brain and the role of transcription factors in lead neurotoxicity. We will examine the interaction between lead and several transcription factors involved in growth and development.

We will particularly focus on POU domain proteins as molecular targets for lead-induced perturbations in transcriptional regulation. We will particularly discuss the influence of lead on the expression and functionality of a member of this family of transcription factors, namely, Oct-2 and its role in mediating the response of CNS neurons to lead exposure.

9.2 TOXIC EFFECTS OF LEAD

The impacts of exposure to neurotoxicants such as lead during development are of special interest because of the sensitivity of the developing nervous system to environmental exposures, and because developmental effects have consequences which can span an individual's entire life (Faustman at al., 2000). The direct neurotoxic actions of lead include apoptosis, excitotoxicity, influence on neurotransmitter storage and release processes, mitochondria, and second messengers. In addition to lead's action on neuronal processes, lead affects cerebrovascular endothelial cells, astroglia, and oligodendroglia. Although all of lead's toxic effects cannot be tied together by a single unifying mechanism, lead's ability to substitute for calcium and zinc is a feature common to many of its toxic actions (Bressler and Goldstein, 1991; Zawia et al., 1998; Lidsky and Schneider, 2003).

Like most other metals lead exists as cations, and as such can react with most ligands present in living cells. Thus, lead has the potential to inhibit enzymes, disrupt cell membranes, damage structural proteins, and affect the genetic code in nucleic acids (Clarkson, 1987). These many potential targets present a great challenge to investigators on mechanisms of action. The effects of lead on the CNS in children range from acute encephalopathy to a chronic subtle change in behavior and cognition. The basic lesions underlying the chronic effects are a matter of great interest and research activity.

9.3 TRANSCRIPTIONAL TARGETS OF LEAD NEUROTOXICITY

It is well known that exposure to lead results in changes in expression of a variety of genes (reviewed by Zawia, 2003). It is important to study gene expression and transcriptional regulation in order to understand the mechanisms associated with lead neurotoxicity. Lead has the ability to interfere with several signal transduction pathways and transcription factors and results in alterations in gene expression. Lead may have its most considerable effect on gene expression patterns via its activation of the calcium, zinc, and phospholipid-dependent protein kinases (Goldstein, 1993; Bressler et al., 1999). The specific effects of lead on gene expression may be due to improper substitution of lead for calcium or zinc on proteins followed by improper expression of "target genes" that are transcriptionally regulated by the lead bound proteins (Zawia and Harry, 1996; Zawia et al., 1998; Bouton and Pevsner, 2000) (Figure 9.1).

Lead can alter patterns of gene expression by the substitution of calcium in specific proteins. Calcium is a second messenger molecule that has specific temporal

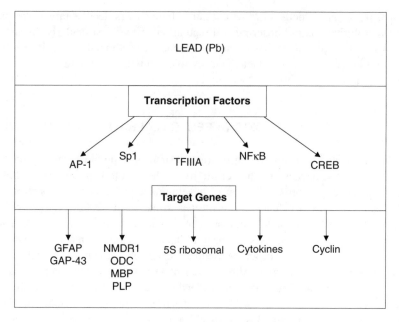

Figure 9.1   Network diagram of the known transcriptional targets for lead. Effects of lead on transcription factor proteins can cause aberrant activity that in turn causes aberrant expression of the protein's target genes expression patterns. Abbreviations used are Sp: specificity protein; CREB: cAMP responsive element-binding protein; AP-1: activator protein-1; NFκB: nuclear-factor-kappa-B; TFIIIA: transcription factor IIIA; GFAP: glial fibrillary acidic protein; GAP-34: growth-associated protein 43; NMDAR-1: *N*-methyl-D-aspartate receptor subunit 1; ODC: ornithine decarboxylase; MBP: myelin basic protein; PLP: myelin proteolipid protein.

and spatial effects on gene expression patterns, which can influence critical biochemical pathways involved in learning and memory in the brain (Ginty, 1997; Hardingham et al., 1997). It is well known that lead may activate protein kinase C (PKC), which is regulated by calcium, and plays a pivotal role in the control of cellular growth and differentiation. Markovac and Goldstein (1988) demonstrated that lead activates PKC at picomolar concentrations, which in turn induces the aberrant phosphorylation of its protein substrates, resulting in disruption of normal cellular functions. The induction of the immediate early genes c-fos, c-jun, and egr-1 by lead in PC12 cells also requires PKC (Kim et al., 1997, 2000). Lead-induced expression of AP-1 transcription factors via PKC is probably responsible for the alteration in the expression of large numbers of target genes that contain AP-1 promoter elements such as glial fibrillary acidic protein (GFAP) (Harry at al., 1996; Zawia and Harry, 1996), growth-associated protein 43 (GAP-43) (Harry at al., 1996; Schmitt et al., 1996; Zawia and Harry, 1996; Pennypacker et al., 1997) and vascular endothelial growth factor (VEGF) (Hossain et al., 2000). One example of the activation of cellular stress response by lead is the activation of NFκB. Ramesh et al. (1999) observed an increase in expression of NFκB in response to lead treatment in PC12 cells through the activation of the MAP kinase family of kinases.

The substitution of lead for zinc in zinc finger protein (ZFP) may also have significant effects on the *de novo* expression of the bound protein as well as any genes that are regulated by that protein (Zawia et al., 1998). The binding of lead to the Sp1 transcription factor results in alterations in the expression of target genes which are regulated by Sp1, such as myelin basic protein (MBP), proteolipid protein (PLP), ornithine decarboxylase (ODC) (Zawia et al., 1994; Zawia and Harry, 1995; Zawia et al., 1998), and the NMDA-R1 receptor subunit (Zawia et al., 1998; Basha et al., 2003). Lead has also been shown to interfere selectively with the DNA-binding properties of Sp1 and TFIIIA, at μ molar concentration (2.5 μM), by acting at the zinc-binding site of these proteins (Zawia et al., 1998; Hanas et al., 1999). Lead may target various transcription factors through different mechanisms. In the case of Sp1, lead can directly compete with zinc in the zinc finger domain of Sp1 or indirectly by interfering with signal transduction/transcription coupling pathways. (For more details, see Chapter 3.)

9.4 POU FAMILY OF TRANSCRIPTION FACTORS IN NEURAL DEVELOPMENT

The actions of lead on ZFP transcription factors have been well characterized (see Chapter 3); however, other genes that are not regulated by ZFP are also modulated by lead exposure. Thus, the potential involvement of other transcription factor families is a strong possibility.

The regulation of the large number of genes encoded in the human genome needs highly sophisticated machinery, particularly during differentiation, where cells undergo a change to an overtly specialized cell type during development. Transcription factors are the main players in gene regulatory system. These factors bind to specific DNA sequences in the gene promoter or enhancer and stimulate or repress expression of the gene. A number of different families of transcription factors have been defined in which the members of each family have similar DNA binding domains (for detail see Chapters 1 and 2). However, the developmental effects of lead require one to investigate transcription factors whose primary role is to regulate developmental gene expression. One such family of transcription factors is the Pit-Oct-Unc (POU) family, originally named for its founder members Pit-1, Oct-1 and Oct-2, and *unc-86*, but since shown to contain a large number of different transcription factors present in both vertebrates and invertebrates.

The genes encoding the mammalian transcription factors Pit-1, Oct-1, and Oct-2 were cloned and were found to share a common domain, which was also found in the nematodes (*C. elegans*) gene *unc-86* (Latchman, 1990, 1995; Latchman et al., 1992). The POU domain common to these factors constitutes the DNA binding domain of the protein and consists of two highly conservative regions, a POU-specific domain (~75 amino acid N-terminal region), which is unique to these factors and a POU homeodomain (60 amino acid C-terminal region), which is related to that found in the homeobox proteins. These two major subdomains are joined by a short flexible linker region.

Homeodomain proteins comprise a superfamily of highly conserved DNA binding factors that are involved in the transcriptional regulation of key developmental processes (Latchman, 1999). Today, there are over 150 entries for POU domains in the SMART database of protein domains involved in signaling events (Schultz et al., 2000). Although this is a mere fraction of the 2000 homeodomain proteins identified so far, the importance of the POU proteins is undeniable, as they are known to regulate many fundamental developmental and homeostatic processes, such as embryogenesis and histone gene expression (Phillips and Luisi, 2000). Thus, for example, the Pit-1 protein plays an essential role in the development of the pituitary gland, and its absence therefore results in dwarfism in both mice and humans. Similarly the *unc-86* mutation in the nematode affects the development of sensory neurons. Moreover, although it was originally thought that the Oct-2 factor was specifically expressed only in B lymphocytes of the immune system, it was subsequently shown by He et al. (1989) to also be expressed in specific neuronal cells (Latchman, 1999).

Homeodomain genes of the POU family are attractive candidates to fill a role as transcriptional regulators of genes activated at puberty. In contrast to the Hox family of homeodomain genes, which is only expressed in the mid- and hindbrain, POU domain genes are widely expressed in the developing forebrain, particularly throughout its ventral aspect (Treacy and Rosenfeld, 1992). It was suggested that some POU domain genes might contribute to regulating specific, differentiated functions of the postnatal neuroendocrine brain due to their persistent expression in discrete neuronal subpopulations of the adult hypothalamus (Alvarez-Bolado et al., 1995). This adult expression appears to be limited to few members of POU domain proteins such as Oct-2 protein (He et al., 1989; Hatzopoulos et al., 1990; Stoykova et al., 1992; Ojeda et al., 1999).

9.5 THE POU DOMAIN: A POTENTIAL NEW TARGET FOR LEAD NEUROTOXICITY

The actions of lead seem to be so diverse that no single target has been implicated as a mediator of the wide range of responses elicited by lead exposure. The study of gene expression and transcriptional regulation is an important aspect of understanding the mechanisms associated with lead neurotoxicity. A possible mechanism through which lead may cause neuronal damage is by perturbation of brain gene expression through alterations in numerous transcription factors and signal transduction intermediates involved in development and differentiation. Although the expression of a number of target genes and the involvement of some transcription factors has been shown to occur following exposure to lead, most of these characterizations were conducted *in vitro*. It is still unknown how many genes and which families of transcription factors are associated with the response to lead exposure. The development of new approaches such as DNA and protein array techniques have provided a new tool to delineate the scope of transcription factors that might be involved in lead neurotoxicity and identify unknown transcription factors that respond to lead exposure. This approach will enhance our understanding of the neurotoxicity of lead and reveal new molecular targets that are altered *in vivo* following lead exposure.

We conducted *in vivo* macroarray analysis in an attempt to identify unknown transcription factors that may be involved in the response to lead exposure. We observed that many transcription factors exhibited distinctive temporal patterns of expression. We also found that the mRNA expression profile of a number of transcriptions factors was altered following lead exposure. The most prominent changes were exhibited by the Sp and Oct families (Figure 9.2).

These two transcription families appear to play a critical role in mediating lead-induced disturbances in developmental gene expression. While the involvement of Sp transcription factors in the response to lead exposure has been previously shown (see Chapter 3), the changes in Oct-1 and Oct-2 expression identify the POU domain as a novel target for lead-induced neurotoxicity.

The identity of the Oct-2 transcription factor initially discovered by macroarray screening analysis was confirmed using RT-PCR techniques. The level of mRNA expression of Oct-2 was significantly elevated on PND 5 in lead exposed animals compared to controls (Figure 9.3). The levels of mRNA expression of both Oct-1 and Sp1, measured by RT-PCR, also follow the same pattern indicating that they may share common activation pathways (results are not shown). Oct-2 mRNA profiling by macroarray techniques and RT-PCR display similar expression profiles, thereby validating our screening analysis (Figure 9.4). The changes in Oct-2 expression identify a novel target for lead suggesting that the POU domain factors may play a critical role in mediating lead-induced disturbances in developmental gene expression.

Changes in the expression profile of a transcription factor do not necessarily imply that the levels of its protein would be altered, consequently elevating its functional activity. Therefore, it is important to relate changes in mRNA expression levels to the DNA-binding properties of a transcription factor. To examine whether lead exposure alters the functionality of Oct-2, we performed DNA-binding assays using the electrophoretic mobility shift assay (EMSA) (Figure 9.5). We found that the DNA-binding of Oct-2 was elevated following lead exposure. The elevation was more pronounced during PND 5 and was reduced on PND 15 and 30. These alterations in DNA-binding may be responsible for mediating lead-induced perturbations in developmental gene expression.

To relate the changes in the biosynthesis of Oct-2 to their DNA-binding activity we compared whether lead-elevated Oct-2 expression corresponds to enhanced DNA-binding activity. Lead-induced developmental patterns of the Oct-2 mRNA (by RT-PCR) and Oct-2 DNA-binding mirror each other, suggesting that the elevations in Oct-2 DNA-binding are a product of *de novo* synthesis (Figure 9.6). These findings indicate that the elevation in Oct-2 mRNA level by lead may be occurring at the transcriptional level.

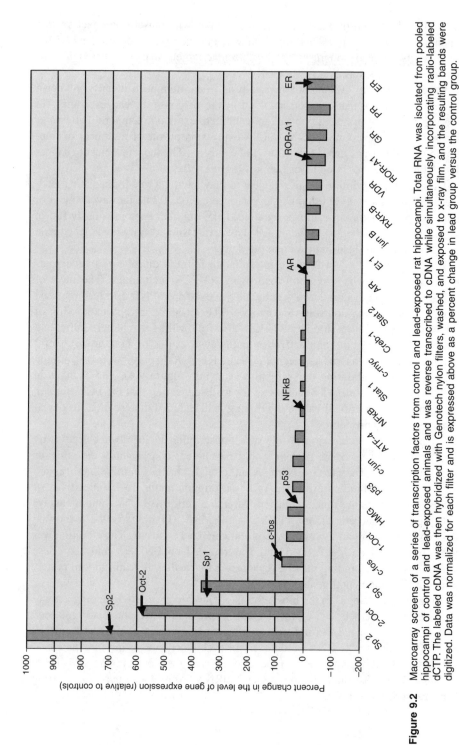

Figure 9.2 Macroarray screens of a series of transcription factors from control and lead-exposed rat hippocampi. Total RNA was isolated from pooled hippocampi of control and lead-exposed animals and was reverse transcribed to cDNA while simultaneously incorporating radio-labeled dCTP. The labeled cDNA was then hybridized with Genotech nylon filters, washed, and exposed to x-ray film, and the resulting bands were digitized. Data was normalized for each filter and is expressed above as a percent change in lead group versus the control group.

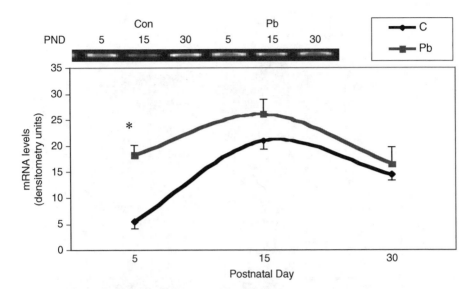

Figure 9.3   Oct-2 mRNA expression (RT-PCR product) in the hippocampus of control and lead-exposed rats. mRNA levels were analyzed using the RT-PCR technique and its corresponding bands were quantitated using image acquisition and analysis software (UVP, Inc., Upland, CA). Values shown are means of three independent experiments. Values indicated with * are significant at $p < 0.05$ over control as evaluated by Student t-test. A representative autoradiogram is shown in the box.

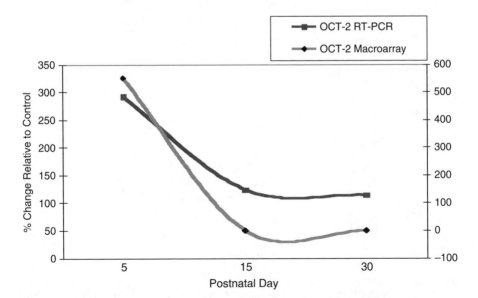

Figure 9.4   Comparison of Oct-2 mRNA profiling by macroarray and RT-PCR following exposure to lead in the rat hippocampus (Oct-2 macroarray in the right axis).

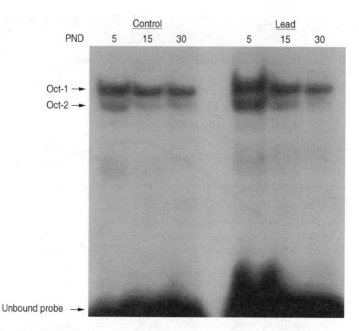

Figure 9.5   The effect of lead on Oct-1 and Oct-2 binding to the octamer motif in the rat hippocampus. Electrophoretic mobility shift assay (EMSA) was performed by using nuclear
extracts derived from the hippocampus tissues of control and lead exposed animals.

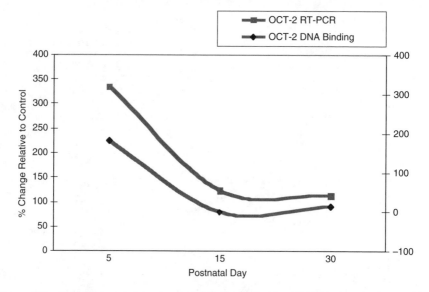

Figure 9.6   The hippocampal developmental profiles of the Oct-2 DNA binding and Oct-2 mRNA
levels (RT-PCR product) following exposure to lead in the rat hippocampus. Oct-2
DNA binding was monitored using EMSA. Shifted bands were scanned and quantitated using image acquisition and analysis software (UVP, Inc., Upland, CA). Oct-2
mRNA level was analyzed using the RT-PCR technique and its corresponding
bands were similarly quantitated. All values shown are expressed as percent
change relative to control (Oct-2 DNA binding in the right axis).

9.6 GENES REGULATED BY OCT-2 AND THEIR TOXICOLOGICAL SIGNIFICANCE

The biological significance of a transcription factor resides in the physiological role of the genes it regulates. Both Oct-1 and Oct-2 target the ubiquitous octamer regulatory element ATGCAAAT, which is present in the promoter and enhancer elements of the immunoglobulin genes (Singh et al., 1986). While Oct-1 is found in all tissue types, Oct-2 is expressed specifically in the immune system's B cells and the neuronal cells, and is absent in other cell types (He et al., 1989). Oct-2 contains an activation domain capable of stimulating gene transcription that is located at the C-terminus of the molecule. In addition, the N-terminus of the molecule contains an inhibitory domain, which is capable of repressing gene transcription (Lillycrop et al., 1994).

The Oct-2 is encoded by a single gene; however, the primary transcript of the Oct-2 gene can be spliced into a number of different protein variants with different activities. This alternative splicing is regulated differently in various cell types. In B lymphocytes, Oct-2.1 is the predominant form produced by alternative splicing. In this form the C-terminal activation domain is present and overcomes the effect of the inhibitory region at the N-terminus enabling Oct-2.1 to stimulate transcription of its target genes. In contrast, in neuronal cells, the predominant forms are Oct-2.4 and Oct-2.5, which lack the C-terminal activation domain, and the presence of the N-terminal inhibitory region therefore results in their having a generally repressive effect on transcription (Lillycrop and Latchman, 1992). Thus, the existence of alternative splicing allows Oct-2 to have either an activating or inhibitory effect on the transcription of its target genes, depending on the cell type involved. This provides an explanation of early observations, which suggested that the Oct-2 binding octamer motif had a generally stimulatory effect on gene expression in B lymphocytes but resulted in the inhibition of gene expression in neuronal cells (Latchman 1995, 1996).

The different roles of Oct-2 protein in the regulation of both early developmental events and differentiated adult brain function is indicated by the widespread neuronal expression of Oct-2 transcripts in the embryonic brain, as opposed to the restricted distribution profile observed in adulthood. In the adult brain, Oct-2 mRNA is present in neurons of only a few regions, including the medial mammillary nucleus and the hypothalamic suprachiasmatic nucleus, the olfactory bulb and tract, the hippocampus and the piriform, insular, and somatosensory regions of the cerebral cortex (Ojeda et al., 1999). These temporal distributions are consistent with our findings following lead exposure. The pronounced elevation in Oct-2 expression on PND 5 suggests that lead may interfere with early postnatal developmental events regulated by POU domain factors. This is further supported by the findings that knockout mice lacking Oct-2 die shortly after birth, which indicates that Oct-2 is essential for postnatal survival (Corocoran et al., 1993). The defects in these mice, which caused their death, such as a failure to feed properly, are more suggestive of deficiencies in the development of the nervous system than in the immune system. This further indicates that Oct-2 likely plays a critical role in the developing nervous system (Latchman, 1999).

The primary role of a transcription factor is to regulate the expression of specific target genes. In the case of Oct-2, several target genes for the inhibitory effect of Oct-2 have been identified in neuronal cells. Thus, Oct-2 can inhibit expression of the herpes simplex virus immediate early genes, which contain the octamer-related TAATGARAT motif in their promoters (Lillycrop et al., 1991). Several recent findings have provided direct evidence supporting the concept that this POU domain protein functions as a transcriptional regulator of either neuropeptide or neurotransmitter genes expressed in the postnatal hypothalamus. Oct-2 has been shown to repress the cellular tyrosine hydroxylase gene promoter in neuronal cells (Dawson et al., 1994; Deans et al., 1995). Deans et al. (1996) also identified the neural nitric oxide synthase (nNOS) gene as the first example of a neuronal gene that is positively regulated by Oct-2 and suggest that the Oct-2 transcription factor may play an important role in regulating the vital function of nNOS within the nervous system. Moreover, the same lab showed that different synaptic vesicle proteins are differentially regulated by the Oct-2 transcription factor. While SNAP-25 is indirectly repressed possibly at a post-transcriptional level, the synapsin I promoter appears to be unique in being directly repressed by Oct-2.4 and Oct-2.5 (Deans et al., 1997). The demonstration that Oct-2 proteins can *trans*-activate the gene encoding nitric oxide synthase, but repress the tyrosine hydroxylase gene, suggests that an action of Oct-2 in neurons may be to regulate the synthesis of particular neurotransmitters.

Lead may affect neurotransmission by interference with neurotransmitters (Shellenberger, 1984). The action of lead may involve interruption in signal transduction mechanisms as well as through disruption in the biosynthesis of neurotransmitters. Several lines of evidence suggest that alteration in the expression of neurotransmitter phenotype-specific genes could play a role in long-term neurobehavioral disturbances associated with lead exposure during early development (Tian et al., 2000). Lead may cause variable changes in several neurotransmitter systems: the dopaminergic and glutamergic systems and especially the NMDA receptor complex (see Chapter 5), which play a role in learning and memory process. These systems are affected by chronic lead exposure (Finkelstein et al., 1998). It is well known that lead has the ability to interact with specific molecular targets involved in neurotransmitter systems. The molecular basis for the effects of lead on neurotransmitter systems remains poorly understood. Our results suggest that Oct-2 is a molecular target of lead, and may thus mediate some of the effects of lead on neurotransmission by disruption in the biosynthesis of key synthetic enzymes involved in the production of neurotransmitters. Studies have shown that subchronic exposure to low levels of lead resulted in region-specific alterations in tyrosine hydroxylase activity in rats exposed to lead. This protein is a key regulatory enzyme in biosynthesis of dopamine (Ramesh and Jadhav, 1998). Moreover, perinatal lead exposure alters the expression of nNOS in rat brain (Chetty et al., 2001), the enzyme responsible for nitric oxide synthesis in the brain. Nitric oxide plays a key role in morphogenesis, synaptic plasticity, and regulates neurotransmitter release. The ability of lead to reduce the expression of NOS and any interference for NOS enzymatic activity could suppress the NO production results in alteration in its functional roles which might effect synaptic development and plasticity at brain regions (Selvin-Testa et al., 1997).

Finally, our data have revealed that the Oct family of transcription factors may be potential targets for lead neurotoxicity. Further work is required to examine target genes whose expression is regulated by Oct-2 in neuronal cells along with the signaling pathways involved following lead exposure to elucidate the critical role of the Oct-2 transcription factor in the regulation of gene expression in neuronal cells and to determine its role in lead neurotoxicity.

REFERENCES

Alvarez-Bolado, G., Rosenfeld, M.G., and Swanson, L.W. (1995), Model of forebrain region-alization based on spatiotemporal patterns of POU-III homeobox gene expression, birthdates, and morphological features, *J. Comp. Neurol.* 355, 237–295.

ATSDR (1993), *Toxicological Profile for Lead*, Agency for Toxic Substance and Disease Registry, TP-92/12, Atlanta, GA.

Basha, M.R. et al. (2003), Lead-induced developmental perturbations in hippocampal Sp1 DNA-binding are prevented by zinc supplementation: in vivo evidence for Pb and Zn competition, *Int. J. Dev. Neurosci.* 21, 1–12.

Bellinger, D., Leviton, A., and Waternaux, C. (1987), Longitudinal analysis of prenatal lead exposure and early cognitive development, *N. Engl. J. Med.* 316, 1037–1043.

Bouton, C.M.L.S., and Pevsner, J. (2000), Effects of lead on gene expression, *NeuroToxicology* 21, 1045–1056.

Bressler, J.P., and Goldstein, G.W. (1991), Mechanisms of lead neurotoxicity, *Biochem. Pharmacol.* 41, 479–484.

Bressler, J.P. et al. (1999), Molecular mechanisms of lead neurotoxicity, *Neurochem. Res.* 24, 595–600.

Byers, R.K., and Lord, E.E. (1943), Late effects of lead poisoning on mental development, *Am. J. Dis. Child.* 66, 471–483.

Canfield, R.L. et al. (2003), Intellectual impairment in children with blood lead concentrations below 10 microg. per deciliter, *N. Engl. J. Med.* 348, 1517–1526.

Centers for Disease Control (1991), *Preventing Lead Poisoning in Young Children: A Statement from the Centers for Disease Control*, Atlanta, GA.

Centers for Disease Control (1992), Surveillance of children's blood lead levels—United States, 1991, *Morbid. Mortal. Weekly Rep.* 41, 620–622.

Chetty C.S. et al. (2001), Perinatal lead exposure alters the expression of neuronal nitric oxide synthase in rat brain, *Int. J. Toxicol.* 20, 113–20.

Clarkson, T.W. (1987), Metal toxicity in the central nervous system, *Environ. Health Perspect.* 75, 59–64.

Corcoran, L.M. et al. (1993), Oct-2, although not required for early B-cell development, is critical for later B-cell maturation and for postnatal survival, *Genes Dev.* 7, 570–582.

Dawson, S.J. et al. (1994), The Oct-2 transcription factor represses tyrosine hydroxylase expression via the heptamer motif in the promoter, *Nucl. Acids Res.* 22, 1023–1028.

Deans, Z. et al. (1995), Direct evidence that the POU family transcription factor Oct-2 represses the cellular tyrosine hydroxylase gene in neuronal cells, *J. Mol. Neurosci.* 6, 159–167.

Deans, Z. et al. (1996), Differential regulation of the two neuronal nitric oxide synthase gene promoters by the Oct-2 transcription factor, *J. Biol. Chem.* 271, 32153–32158.

Deans, Z.C. et al. (1997), Differential regulation of genes encoding synaptic proteins by the Oct-2 transcription factor, *Brain Res. Mol. Brain Res.* 51, 1–7.

Duggan, M.J., and Inskip, M.J. (1985), Childhood exposure to lead in surface dust and soil: a community health problem, *Public Health Rev.* 13, 1–54.

Faustman, E.M. et al. (2000), Mechanisms underlying children's susceptibility to environmental toxicants, *Environ. Health Perspect.* 108, 3–21.

Finkelstein, Y., Markowitz, M.E., Rosen, J.F. (1998), Low-level lead-induced neurotoxicology in children: an update on central nervous system effects, *Brain Res. Rev.* 27, 168–176.

Florini, K.L., Krumbhaar, G.D., and Silbergeld, E.K. (1990), *Legacy of Lead: America's Continuing Epidemic of Childhood Lead Poisoning. A Report and Proposal for Legislative Action,* Environmental Defense Fund, Washington, DC.

Ginty, D.G. (1997), Calcium regulation of gene expression: isn't that spatial? *Neuron* 18, 183–186.

Goldstein, G.W. (1990), Lead poisoning and brain cell function. *Environ. Health Perspect.* 89, 91–94.

Goldstein, G.W. (1992), Neurologic concepts of lead poisoning in children, *Pediatric Ann.* 21, 384–388.

Goldstein, G.W. (1993), Evidence that lead acts as a calcium substitute in second messenger metabolism, *NeuroToxicology* 14, 97–101.

Goyer, R.A. (1996), Toxic effects of metals, *Cassaret and Doull's Toxicology: The Science of Poisons,* McGraw-Hill, New York, 643–690.

Hanas, J.S. et al. (1999), Lead inhibition of DNA-binding mechanism of Cys(2)His(2) zinc finger proteins, *Mol. Pharmacol.* 56, 982–988.

Hardingham, G.E. et al. (1997), Distinct functions of nuclear and cytoplasmic calcium in the control of gene expression, *Nature* 385, 260–265.

Harry, G.J. et al. (1996), Lead-induced alterations of glial fibrillary acidic protein (GFAP) in the developing rat brain, *Toxicol. Appl. Pharmacol.* 139, 84–93.

Hatzopoulos, A.K. et al. (1990), Structure and expression of the mouse Oct2a and Oct2b, two differentially spliced products of the same gene, *Development* 109, 349–362.

He, X. et al. (1989), Expression of a large family of POU domain regulatory genes in mammalian brain development, *Nature* 340, 35–42.

Hossain, M.A. et al. (2000), Induction of vascular endothelial growth factor in human astrocytes by lead, *J. Biol. Chem.* 276, 27874–27882.

Jacobs, D.E. et al. (2002), The prevalence of lead-based paint hazards in U.S. housing, *Environ. Health Perspect.* 110, A599–A606.

Kim, K. et al. (1997), Induction of c-fos rnRNA by lead in PCI2 cells, *Int. J. Dev. Neurosci.* 15, 175–182.

Kim, K. et al. (2000), Immediate early gene expression in PC12 cells exposed to lead: requirement for protein kinase C, *Neurochemistry* 74, 1140–1146.

Lanphear, B.P. et al. (2000), Cognitive deficits associated with blood lead concentrations <10 microg/dl in US children and adolescents, *Public Health Res.* 115, 521–529.

Latchman, D.S. (1990), *Gene Regulation: Eukaryotic Perspective,* Unwin Hyman, London.

Latchman, D.S. et al. (1992), POU family transcription factors in sensory neurons, *Biochim. Soc. Trans.* 20, 627–631.

Latchman, D.S. (1995), *Eukaryotic Transcription Factors,* 2nd ed., Academic Press, London.

Latchman, D.S. (1996), The Oct-2 transcription factor, *Int. J. Biochem. Cell Biol.* 28, 1081–1083.

Latchman, D.S. (1999), POU family transcription factors in the nervous system, *J. Cell. Physiol.* 179, 126–133.

Lidsky, T.I., and Schneider, J.S. (2003), Lead neurotoxicity in children: basic mechanisms and clinical correlates, *Brain* 126, 5–19.

Lillycrop, K.A. et al. (1991), The octamer binding protein Oct-2 represses HSV immediate early genes in cell lines derived from latently infectable sensory neurons, *Neuron* 7, 381–390.

Lillycrop, K.A., and Latchman, D.S. (1992), Alternative splicing of the Oct-2 transcription factor is differentially regulated in B cells and neuronal cells and results in protein isoforms with opposite effects on the activity of octamer/TAATGARAT-containing promoters, *J. Biol. Chem.* 267, 24960–24966.

Lillycrop, K.A. et al. (1994), Repression of a herpes simplex virus immediate-early promoter by the Oct-2 transcription factor is dependent upon an inhibitory region at the N-terminus of the protein, *Mol. Cell. Biol.* 14, 7633–7642.

Markovac, J., and Goldstein, G.W. (1988), Lead activates protein kinase C in immature rat brain microvessels, *Toxicol. Appl. Pharmacol.* 96, 14–23.

Needleman, H.L. (1994), Childhood lead poisoning, *Curr. Opin. Neurol.* 7, 187–190.

Needleman, H.L. (1998), Childhood lead poisoning: the promise and abandonment of primary prevention, *Am. J. Public Health* 88, 1871–1877.

Ojeda, S.R. et al. (1999), The Oct-2 POU domain gene in the neuroendocrine brain: a transcriptional regulator of mammalian puberty, *Endocrinology* 140, 3774–3789.

Pennypacker, K.R. et al. (1997), Lead-induced developmental changes in AP-1 DNA binding in rat brain, *Int. J. Dev. Neurosci.* 15, 321–328.

Perlstein, M.A., and Attala, R. (1966), Neurological sequelae of plumbism in children, *Clin. Ped.* 5, 292–298.

Phillips, K., and Luisi, B. (2000), The virtuoso of versatility: POU proteins that flex to fit, *J. Mol. Biol.* 302, 1023–1039.

Pirkle, J.L. et al. (1994), The decline in blood lead levels in the United States: the National Health and Nutrition Examination Surveys, *J. Am. Med. Assoc.* 272, 284–291.

Ramesh, G.T. and Jadhav, A.L. (1998), Region-specifle alterations in tyrosine hydroxylase activity in rats exposed to lead, *Mol. Cell. Biochem.* 189, 19–24.

Ramesh, G.J. et al. (1999), Lead activates Nuclear Transcription Factor-kB, Activator Protein-1, and amino-terminal c-Jun kinase in Pheochromacytoma cell, *Toxicol. Appl. Pharmacol.* 155, 280–286.

Schmitt, T.J., Zawia, N., and Harry, G.J. (1996), GAP-43 mRNA expression in the developing rat brain: alterations following lead-acetate exposure, *NeuroToxicology* 17, 407–414.

Schultz, J. et al. (2000), SMART: A web-based tool for the study of genetically mobile domains, *Nucleic Acids Res.* 28, 231–234.

Selvin-Testa, A. et al. (1997), The nitric oxide synthase expression of rat cortical and hippocampal neurons changes after early lead exposure, *Neurosci. Lett.* 236, 75–78.

Shellenberger M.K. (1984), Effects of early lead exposure on neurotransmitter systems in the brain. A review with commentary, *NeuroToxicology* 5, 177–212.

Singh, H. et al. (1986), A nuclear factor that binds to a conserved sequence motif in transcriptional control elements of immunoglobulin genes, *Nature* 319, 154–158.

Stoykova, A.S. et al. (1992), Mini-Oct and Oct-2c: two novel, functionally diverse murine Oct-2 gene products are differentially expressed in the CNS, *Neuron* 8, 541–558.

Treacy, M.N., and Rosenfeld, M.G. (1992), Expression of a family of POU-domain protein regulatory genes during development of the central nervous system, *Annu. Rev. Neurosci.* 15, 139–165.

Tian, X., Sun, X., and Suszkiw, J.B. (2000), Upregulation of tyrosine hydroxylase and downregulation of choline acetyltransferase in lead-exposed PC12 cells: the role of PKC activation, *Toxicol. Appl. Pharmacol.* 167, 246–252.

World Health Organization (WHO) (1994), *Environmental Health Criteria on Inorganic Lead*, Geneva.

Zawia, N.H., Evers, L.B., and Harry, G.J. (1994), Developmental profiles of ornithine decarboxylase activity in the hippocampus, neocortex and cerebellum: modulation following lead exposure, *Int. J. Dev. Neurosci.* 12, 25–30.

Zawia, N.H., and Harry, G.J. (1995), Exposure to lead acetate modulates the developmental expression of myelin genes in the rat frontal lobe, *Int. J. Dev. Neurosci.* 13, 639–644.

Zawia, N.H., and Harry, G.J. (1996), Developmental exposure to lead interferes with glial and neuronal differential gene expression in the rat cerebellum, *Toxicol. Appl. Pharmacol.* 138, 43–47.

Zawia, N.H. et al. (1998), SP1 as a target site for metal-induced perturbations of transcriptional regulation of developmental brain gene expression, *Dev. Brain Res.* 107, 291–298.

Zawia, N.H. (2003), Transcriptional involvement in neurotoxicity, *Toxicol. Appl. Pharmacol.* 190, 177–188.

CHAPTER 10

Ethanol Neurotoxicity: Expanding Horizons from Stem Cells to Microchips

Mohan C. Vemuri and Rajgopal Yadavalli

CONTENTS

10.1 ETHANOL AND THE CENTRAL NERVOUS SYSTEM: AN OVERVIEW

Alcohol is a neuroactive addictive depressant drug, and chronic ethanol consumption has been associated with pleiotropic effects on cellular function. The mechanism by which it exerts its toxic effects is still unclear. Ethanol is known to elicit neurotoxicity characterized by ataxia, loss of coordination, hyperexcitation, convulsions, and paralysis. Excess alcohol consumption alters neuronal growth and causes striking neuronal degeneration and cell death in several brain regions, resulting in loss of memory and cognitive impairments. Several mechanisms, such as calcium overload, excitotoxicity, free radical injury, and disturbance in protein synthesis have been implicated in chronic ethanol-mediated neuronal cell death. By contrast, the signal transduction pathways that converge in cell death are incompletely defined and briefly reviewed in this chapter in the context of ethanol-induced neurotoxicity. Further, applications and utility of recent advances in microarrays,

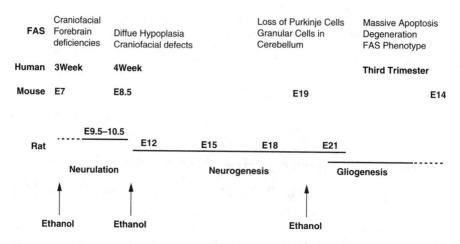

Figure 10.1   Selective vulnerability of embryonic cell populations to ethanol exposure. A schematic comparison across rat, mouse, and human development. Ethanol exposure
at different stages of embryonic development results in specific fetal alcohol syndrome phenotype deficiencies both in animals and humans.

gene expression profiles, transgenic knockout technology, and stem cell biology have
been considered in alcoholic neuropathology.

Chronic alcohol abuse results in neurological diseases such as Korsakoff's syndrome, Marchiafava-Bignami disease, pellagrous encephalopathy, and acquired
hepatocerebral degeneration (Charness, 1993; Victor, 1994). However, these diseases
are not due to alcohol per se, but to nutritional deficits that occur in chronic alcoholics
(Victor, 1994). Excess alcohol consumption causes neuronal degeneration in different brain regions, resulting in memory loss and cognitive impairments (Arendt et al.,
1988). A particularly notable paradigm of ethanol effects on development includes
fetal alcohol syndrome (FAS), which is an abnormal *in utero* growth of the fetus
due to maternal alcohol consumption. FAS children are characterized by mental
retardation, microcephaly, hyperactivity, intellectual deficits, decreased intelligence
quotients, motor abnormalities, and other behavioral problems. It is well established
now that the developing brain is highly vulnerable to the effects of ethanol, causing
disruption in central nervous system (CNS) development including decreased neurogenesis, delayed or aberrant neuronal migration, anomalous morphological development (Miller, 1993), changes in the ontogeny of neurotransmitter synthesis
(Pellegrino and Druse, 1992), and neuronal depletion in several brain regions. All
these changes contribute to devastating developmental deficits in the CNS of experimental animal models of fetal alcohol exposure (FAE) and FAS children (Pellegrino
and Druse, 1992).

FAS features are also reflected in animal models and are sustained throughout
the growth, exhibiting behavioral and biochemical abnormalities (Figure 10.1). The
cellular mechanisms by which ethanol induces damage *in utero* are not well understood. However, oxidative stress is believed to be one of the putative mechanisms.

Ethanol-induced oxidative damage to the fetus could be attenuated by a variety
of antioxidants as was documented in whole animal and tissue culture studies
(Cohen-Kerem and Koren, 2003). Several reviews have dealt with the mechanisms

of ethanol fetotoxicity in detail (Michaelis, 1990; Olney et al., 2002b). In general, all agree that the fetus is more susceptible to ethanol than the mother, and offsprings born to alcoholic mothers have growth retardation and oxidant stress, and may suffer mutagenicity. Failure in the transport of nutrients, hypoxia, and calcium handling mechanisms such as N-methyl-D-aspartate (NMDA) receptors are part of the cellular injury by ethanol in the fetus. Selective vulnerability of embryonic cell populations to cell death due to high caspase-3 activity has also been implicated (Ikonomidou et al., 2000; Olney et al., 2002a), particularly when the cells are exposed to ethanol during early, mid, or late stages of CNS growth and development, when cells are just beginning to form synaptic connections. Animal and cell culture studies indicate that use of vitamin E and β-carotene in hippocampal cultures (Mitchell et al., 1999a; Mitchell et al., 1999b), flavonoids, vitamin E, folic acid, and β-carotene in animal studies (Tanaka et al., 1988) show varying levels of protection in FAS conditions. The available experimental evidence suggests that the use of antioxidants in alcohol-consuming mothers and the use of NMDA antagonist lessens the apoptotic burden. However, use of these agents in clinical settings has not been tried yet and should be seriously considered as a means to reduce fetal alcohol damage.

10.2 THE NEUROTOXICITY OF ETHANOL: FROM CELLS TO WHOLE ANIMALS

Neuronal cell cultures exposed to ethanol provided a model for examining the direct toxic effects of ethanol. Acute and chronic ethanol treatment alters the function of neurotransmitter receptors, ion channels, and transport processes (Tabakoff et al., 1979). At low concentration, ethanol prevents NMDA-activated currents in primary neuronal cultures. In dissociated brain cells, ethanol inhibits NMDA-stimulated increases in intracellular calcium concentrations (Dildy and Leslie, 1989). Ethanol also inhibits the release of ^{3}H-noradrenaline from brain cortex (Fink and Gothert, 1990) and endogenous dopamine from striatal slices (Woodward and Gonzales, 1990), in response to NMDA stimulation. Studies using cultured cerebellar neurons have found that chronic ethanol treatment can significantly increase NMDA-stimulated calcium influx (Iorio et al., 1992) that would sensitize neurons to excitotoxicity. Chronic ethanol treatment of cells *in vitro* in cultures and *in vivo* in animals results in increased dihydropyridine-sensitive voltage-dependent calcium channels (Brennan et al., 1990). In addition, ethanol can directly increase $[Ca^{2+}]i$ by releasing intracellular stores (Daniell et al., 1987; Machu et al., 1989). Ethanol induced inhibition of NMDA receptor function and increased intracellular Ca^{2+} concentration $[Ca^{2+}]i$ in cerebellar granule neurons grown in a medium containing 5 mM KCl caused apoptotic neuronal death (Bhave and Hoffman, 1997). In addition, chronic ethanol increases receptor-stimulated production of nitric oxide, which in turn is known to increase excitotoxicity (Crews et al., 1998). Thus, several studies suggest that chronic ethanol treatment may disrupt calcium homeostasis causing excitotoxicity. Ethanol is also shown to interfere with glutamate metabolism due to decreased glutamine synthesis in astrocyte cultures leading to an increased sensitivity to withdrawal seizures by NMDA. Chronic ethanol treatment decreased the efficiency of GABAergic transmission in synaptosomes (Morrow et al., 1990), and decreases GABA-stimulated chloride flux (Harris and Allan, 1989). In addition to receptor-gated channels, ethanol exposure increased the maximum velocity of Na^+/Ca^{2+} antiporter (Michaelis, 1990). Alterations in dopaminergic and muscarinic cholinergic receptors after chronic ethanol exposure may be the result of ethanol-induced impairment of receptor effector coupling. Ethanol (5 mM) inhibits long term potentiation (LTP) in rat hippocampal neurons, providing a biological correlate of ethanol induced memory impairments (Blitzer et al., 1990). Since neurotransmission is primarily affected (involving different neurotransmitters and their receptors), the ultimate action of ethanol leading to cognitive impairment may mostly lie in the signal transduction process.

Exposure to ethanol during development causes well-characterized decreases in the number of both cerebellar granule cells and Purkinje neurons and the neurotoxicity appears to occur at specific developmental times of vulnerability (Hamre and West, 1993; Napper and West, 1995). Addition of ethanol to a fresh cerebellar cell culture (less than 3 days) increased cell loss (Luo and Miller, 1997). However, with continued time in culture, the vulnerability of the granule cells to the neurotoxicity of ethanol was diminished. Thus, in 7 to 8-day-old cultures, a 24-hour treatment with ethanol was reported not to alter granule cell survival (Castoldi et al., 1998; Zhang et al., 1998). In the neuronal-like PC12 cell line, ethanol alone did not alter cell viability, but induced apoptosis in the presence of permissive conditions (Oberdoerster et al., 1998). Similarly, when granule cell cultures were incubated under nondepolarizing conditions, ethanol was reported to enhance granule cell death by increased apoptosis (Wegelius and Korpi, 1995; Bhave and Hoffman, 1997). In contrast, others failed to observe an ethanol-mediated increase in granule cell death under nondepolarizing conditions (Castoldi et al., 1998; Zhang et al., 1998). However, numerous methodological differences exist that can complicate the search for the precise neurotoxic effects of ethanol on granule cells.

Alcohol treated animal studies exhibit changes including a reduction in brain volume that is comparable to the observed decrease in human brain volume (Harper and Kril, 1991). This is mainly due to the loss of neurons in specific regions of the brain in chronic human alcoholics, a pathological process known as alcohol-related neuronal loss (ARNL), responsible for *alcoholic dementia* or *dementia associated with alcoholism* (DAA). Brain shrinkage is due to loss of white matter secondary to neurodegeneration (Harper and Kril, 1991; Harper, 1998). Neuroimaging studies have demonstrated a progressive decrease in the volume of gray matter in the frontal lobes of chronic alcoholics, which is superimposed on a gradual decrease in the volume of frontal cortical gray matter seen during normal aging (Pfefferbaum et al., 1997). Neuroimaging studies confirmed that volume reductions in gray matter are partially reversible with alcohol abstinence (Pfefferbaum et al., 1998). Loss of gray matter in the human brain may be due to loss of neurons, shrinkage of neuronal cell bodies, or reduction in the number and extent of dendrites. Animal studies suggest the possibility of a combination of atrophic cell death of a subset of neurons accompanied by sprouting of new connections in surviving neurons. Arendt et al. carried out a series of studies on the effect of long-term exposure to ethanol on basal forebrain neurons in rats, and detected a complex response in which some neurons in the basal forebrain of ethanol drinking rats are lost, but the surviving neurons undergo a remodeling process characterized by increased dendritic complexity (Arendt et al., 1995). Careful neuropathological analyses have provided evidence for both the loss of neurons as well as shrinkage of individual neurons in chronic alcoholics (Kril et al., 1997; Kril and Halliday, 1999). Interestingly, neuronal loss is seen in certain regions of the cortex, e.g., the frontal association cortex, while other regions such as the motor cortex are unaffected. A combined analysis of several cortical regions may be the reason for other investigators not able to detect neuronal loss in the brains of alcoholics (Jensen and Pakkenberg, 1993). Brain regions with histological neuronal loss correspond to the regions where neuroimaging studies demonstrate the largest decreases in gray matter volume, e.g., the frontal cortical regions (Kril and Halliday, 1999).

10.3 MECHANISMS OF ETHANOL NEUROTOXICITY: APOPTOSIS VERSUS NECROSIS

The mechanisms by which ethanol exerts its toxic effects are still unclear, particularly aspects such as tissue damage secondary to cellular injury, carcinogenesis, and teratogenesis. The specific actions of ethanol are related to its ability to interact with neuronal membranes leading to changes in membrane fluidity, neurotransmitter action, and signal transduction (Goldstein and Chin, 1981). A number of mechanisms contributing to ethanol-related neuropathology have been suggested (Figure 10.2). These include induction of hypoxic, hypoglycemic, or ischemic fetal environment, possibly as a consequence of decreased umbilical circulation, and enhanced production of free radicals together with diminished production of endogenous protective antioxidants (Mantle and Preedy, 1999). Changes in calcium homeostasis due to ethanol favor cellular damage or death, while alterations in cell metabolism due to DNA hypomethylation, or secondary metabolic damage due to acetaldehyde, also severely hamper the metabolic status of cells (Hillbom, 1999). The cytotoxic effects of ethanol are not entirely clear, but certain possibilities are suggested.

First, ethanol can exert cytotoxic effects through its lipid-soluble properties and might display its biological effects by physical action as a denaturing agent of macromolecules. Second, the toxic effects of ethanol are linked to its metabolic fate and probably are mediated by oxygen-dependent generation of free radicals, which may directly react with vital cellular constituents, and may even be transformed into more reactive species. Third, the cytotoxic effects of ethanol may result from a combined influence of its physical, chemical and metabolic properties (Diamond and Gordon, 1997). Recent research has suggested an intriguing possibility that ethanol may exert its cell toxicity via DNA damage (Brooks, 2000). The DNA damage by ethanol and its major metabolite acetaldehyde may be mediated by generation of free radicals. Several studies showed the involvement of free radical induced damage to chromatin and proteins in the brain and liver after chronic ethanol exposure (Puntarulo and Cederbaum, 1989; Mahadev and Vemuri, 1998a). Further,

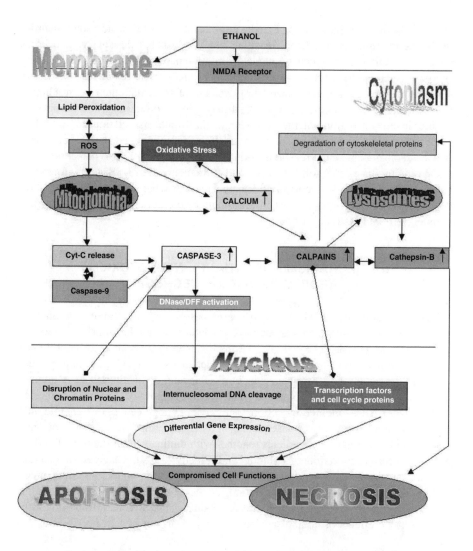

Figure 10.2 Schematic representation of mechanisms of ethanol neurotoxicity. Ethanol affects various molecules in all three compartments of the cell, namely membranes, cytoplasm, and nucleus. The cellular events finally converge into cell death as described in detail in the text, due to ethanol exposure *in utero* as well as *in vivo* in animals and in cellular models of alcoholism.

ethanol is reported to induce oxidative stress in brain through its lipophilic and free radical-generating properties (Renis et al., 1996). Cell injury caused by the effects of oxidative stress has led to a number of interesting working hypotheses that link cellular redox balance to induction of apoptosis (Kroemer et al., 1995). Supporting such a possibility, a role for apoptosis-related gene expression in alcohol-induced brain damage was proposed (Freund, 1994).

Ethanol-mediated reductions in neuronal number could be due to increased cell death. Apart from apoptosis, necrosis has also been implicated in ethanol-induced decrease in cell numbers (Zhang et al., 1998). Appearance of pyknotic nuclei and morphological evidence of apoptosis in the brains of fetal mice exposed to ethanol was also reported (Bannigan and Burke, 1982). These studies indicate that ethanol-induced neuronal loss *in vivo* involves increased apoptosis. Although the reasons for such reduced neuronal numbers remain elusive, ethanol-induced neuronal cell loss could arise due to decreased cell proliferation, cell migration apart from increased cell death during development (Marcussen et al., 1994; Diamond and Gordon, 1997). Compounding the ability of ethanol to elicit cell death, neurotrophic factors are also implicated (MacLennan et al., 1995). Ethanol decreased the expression of low affinity neurotrophin receptor and p75 (Luo and Miller, 1997; Luo et al., 1997). However, ethanol-induced apoptosis is not consistent with a reduction in p75 levels (Barrett and Georgiou, 1996). In contrast, induction of apoptosis by ethanol might involve the neurotrophin tyrosine kinase receptor, trkA, which promotes cell survival (Fagan et al., 1996). In addition, ethanol inhibits tyrosine kinase activity of the insulin-like growth factor-1 receptor in C6 rat glioblastoma cells and Balb/c 3T3 fibroblasts (Resnicoff et al., 1994). It is therefore likely that a similar action on neurotrophin signaling could be involved in ethanol-induced apoptosis.

An increased role for apoptosis-related gene expression in ethanol-induced brain damage has also been suggested (Freund, 1994; Hamby-Mason et al., 1997). Recent studies also provide evidence that ethanol induces apoptotic pattern of neurodegeneration in a rodent model of fetal alcohol syndrome (Ikonomidou et al., 2000) and caspases-3 activation in developing mouse brain (Olney et al., 2002a). Ethanol treatment *in vitro* in primary cultures of cerebellar granule cells, hepatocytes, lymphocytes, and thymocytes also seem to favor apoptosis (Bhave and Hoffman, 1997; Zhang et al., 1998). The important implication of all these studies is that the ethanol-induced neuronal loss primarily involves an activated apoptotic mechanism. In this regard, our recent studies showed that chronic ethanol treatment caused a differential modulation of apoptosis-associated proteins, cytochrome *c* release, concomitant with procaspase-9 and procaspase-3 activation, leading to oligonucleosomal DNA fragmentation in rat cerebral cortex and cerebellum (Rajgopal et al., 2003).

Diverse groups of molecules are involved in the apoptosis pathway. Caspases and calpains, which belong to the cysteine protease family, are implicated in the process of apoptosis. Caspases are activated by proteolytic cleavage in response to a number of apoptotic stimuli and participation of caspases has been well documented in various CNS experimental apoptotic models (Marks and Berg, 1999). Calpain or calcium-dependent neutral protease shows an absolute requirement of intracellular free calcium for activity. Relatively little is known about the molecular mechanisms of cell death pathways mediated by caspases and calpains in ethanol-mediated neurotoxic insult, particularly in the *in vivo* experimental paradigm. Our studies show enhanced calpain activity and cleavage of α-spectrin resulting in calpain signature breakdown products, indicating cytoskeletal damage that could converge into cell death (Rajgopal and Vemuri, 2002).

Caspase-3 proform (32-kDa) showed decreased immunoreactivity in cortex and cerebellum, while the cleaved active fragment (17-kDa) increased significantly in the cerebellum after ethanol treatment. Chronic ethanol treatment increased caspase-3 activity in the cortex and to a higher extent in the cerebellum, which was further confirmed by blocking experiments with caspase-3 specific inhibitor, N-acetyl-Asp-Glu-Val-Asp-aldehyde (Ac-DEVD-CHO). We tested whether activated caspase-3 cleaves downstream substrates such as poly (ADP-ribose) polymerase-1 and protein kinase C-delta (PKC-δ). Western blots showed poly (ADP-ribose) polymerase-1 cleaved its signature fragment of 85-kDa and decreased the levels of PKC-δ in the cerebral cortex and the cerebellum after ethanol treatment, suggestive of caspase-3 activation. Elevated caspase-3 activity in the cerebellum than the cortex correlated with cytochrome c, caspase-9, active caspase-3 (p17), poly (ADP-ribose) poly-merase-1, and PKC-δ. This suggests a mechanism by which ethanol might be exerting pro-apoptotic events in the brain and demonstrates that selective brain regions such as cerebellum are vulnerable to ethanol neurotoxicity. Activation of caspase-3 cascade in the rat brain following *in vivo* chronic ethanol treatment might contribute to the process of apoptosis by inducing structural alterations, affecting cell-signaling pathways and DNA repair mechanisms differently in different brain regions, such as the cerebral cortex and the cerebellum.

The apoptotic process is critical in executing physiological and pathological cell death. It is controlled and regulated by extracellular survival signals (Climent et al., 2002) as well as intrinsic intracellular signal pathways mediated by protein phos-phorylation. Ethanol has profound effects on the signal transduction cascades that regulate protein kinase/phosphatase systems (Diamond and Gordon, 1997). It is known that ethanol alters neuronal activity by primarily deranging the neurotransmitter function, ion channel conductivity, RNA metabolism, and protein synthesis. These functions are regulated by protein phosphorylation and dephospho-rylation reactions. Changes in specific protein kinases in ethanol-induced apoptosis suggest that apoptosis cascade might be subject to factors that can potentiate or inhibit the process. Hence, the identification of apoptosis-relevant phosphoproteins and elucidation of their function is of prime importance to advance the understanding of how protein phosphorylation modulates apoptosis. Ethanol changes the redox state and the intracellular calcium levels, which in turn modulate calcium-dependent protein kinases and proteases. It is shown that acute ethanol decreases the calcium uptake while chronic ethanol increases intracellular calcium (Messing et al., 1986). Chronic ethanol treatment increases the levels of PKC and protein kinase mediated phosphorylation in neural cells, implicating PKC in the up-regulation of Ca^{2+} chan-nels (Messing et al., 1990; Mahadev and Vemuri, 1998b). Ethanol induced much higher Ca^{2+} uptake in the presence of phorbol esters, which suggests that ethanol might alter some other pathways in addition to PKC to enhance Ca^{2+} uptake. Possible mechanisms include alterations in cAMP levels (Gordon et al., 1986) and cAMP-dependent protein phosphorylation. Chronic ethanol exposure reduces adenosine stimulation of cAMP content in cultures (Gordon et al., 1986; Rabin, 1993). This heterologous desensitization is due to decrease in Gsα mRNA and protein, leading to decreased cAMP production (Charness et al., 1988). Studies involving the phos-phorylation of a 50 kDa nuclear protein in G_0/G_1 phase of the cell cycle in glial cultures and in G_0 cells of adult rat cerebrum support the role of phosphorylation mediated cell cycle kinetics by ethanol (Hinson et al., 1991).

The complexity and heterogeneity of regulated cell death and the huge amount of information on protein kinases and protein phosphatases suggest that protein phosphorylation may have multiple roles in the control of cell death. The altered phosphorylation pattern associated with death signaling in apoptotic cells has been termed dysphosphorylation (Robaye et al., 1994). Increased phosphorylation of hitherto unidentified proteins has been reported in apoptosis. The identification of apoptosis-relevant phosphoproteins and elucidation of their function is of critical importance to advance our understanding of how protein phosphorylation modulates apoptosis. Studies from our laboratory and several others have investigated the effect of ethanol on signal transduction and reported changes in endogenous protein phosphorylation, CaM Kinase II activity, PKC activity, and its isoforms βI, βII, and δ-levels in rat brain during prenatal and postnatal ethanol treatment (Diamond and Gordon, 1997; Haviryaji and Vemuri, 1997; Mahadev and Vemuri, 1998a; Mahadev and Vemuri, 1998b). The ability of chronic ethanol treatment on members of the MAP kinase family (viz., extracellular signal-regulated kinase [ERK], c-Jun N-terminal kinase [JNK] and p38) and the role of cyclin-dependent protein kinase (Cdk5), calcium-dependent tyrosine kinase (PYK2), and Akt-kinase was examined recently (Rajgopal et al., 2003). It is an accepted notion that each group of MAPK family members has separate signaling pathways, and that JNK and p38 mediate different, even opposite, events from those by ERK. The two ERK isoforms (ERK1/2) are highly expressed in the normal brain, where they are predominantly located within neurons (Boulton et al., 1991). In neuronal cells, JNK and p38 have been reported to be important in neuronal apoptosis, whereas ERK induces differentiation or cell growth. However, evidence exists that each of the MAPK family members cross talks and converges on some cascade to balance the separate pathways (Su and Karin, 1996). Chronic ethanol treatment activated the three MAPKinases (ERK, JNK, and p38) differentially in the rat cerebral cortex and cerebellum. Increased ERK, JNK, and p38 activity may contribute to neuronal cell death by shifting the balance from protective ERK to the degenerative JNK/p38 signal transduction (Xia et al., 1995). However, the consequences of the activation of the MAPK family in chronic ethanol paradigm are not clear yet. The increase in ERK, JNK, and p38 activity appears to be a part of the neuronal response following chronic ethanol treatment, and these findings differ substantially from some of the *in vitro* experiments reported earlier in many models, and underline the necessity to analyze the neuronal stress pathways in the adult brain. Furthermore, MAPK family members were differentially activated in the cortex and cerebellum, which indicates that there could be some differences in the signaling system for the activation of each MAPK family.

The process of apoptosis is also linked with the regulation of the cell cycle (Ellis et al., 1991). In the mammalian central nervous system, neurons destined to die via apoptosis transiently but abortively re-enter the cell cycle, before proceeding to death (Nuydens et al., 1998). Cell cycle progression is controlled through the ordered expression and activation of a growing number of proteins that include cyclins and their associated cyclin-dependent kinases (Cdks). While Cdks such as Cdk1, Cdk2, Cdk4, and Cdk6 play important roles in cell proliferation, others such as Cdk5 do not appear to be involved in the regulation of the cell cycle. The expression of Cdk5 and its regulators, p35 and p67, was investigated in the adult rat cerebral cortex and cerebellum, using an experimental paradigm of *in vivo* chronic ethanol exposure (Rajgopal and Vemuri, 2001). There were no appreciable changes in the expression of Cdk5 protein levels, while its regulatory proteins, p35 and p67, showed decreased levels following chronic ethanol treatment. However, ethanol treatment resulted in increased Cdk5 activity, but not its levels, in both the cortex and cerebellum with relatively high activity in the cortex. Given the abundant expression and functions of Cdk5 in neural cells, these data support a regulatory role for Cdk5 in ethanol-mediated cell injury and may contribute to impaired CNS development in brain atrophy associated with alcoholic neurodegeneration. Prenatal ethanol exposure caused the down-regulation of p27(Kip) between P0 and P21 and was found to disturb the expression of cell cycle machinery in the postnatal cerebellum. The changes in cyclin-dependent kinases may account for the teratogenic effects of ethanol on the developing cerebellum (Li et al., 2002). Further, ethanol-mediated perturbations in the ontogeny of neurotransmitter synthesis and release may most likely involve changes in the p67 (Munc-18) protein, given the association of p67 with neurosecretion. Furthermore, ethanol-induced neuronal cell death may involve Cdk5/p35 activation, probably through regulating the phosphorylation of neuronal cytoskeletal proteins, since increased Cdk5 activity has been reported in apoptotic cell death (Namura et al., 1998).

Chronic ethanol treatment results in increased intracellular calcium in neural cells, and a sustained high intracellular free Ca^{2+} may initiate a cascade of signals leading to activation of phospholipases, endonucleases, and calcium-dependent proteases. Calcium-dependent tyrosine kinase, also known as PYK2/RAFTK, is activated by a variety of extracellular stimuli that elevate cytosolic calcium concentration (Lev et al., 1995; Dikic et al., 1996). Activation of PYK2 plays a role in regulating neurotransmission and neuroplasticity by phosphorylating potassium channels (Lev et al., 1995) and apoptotic cell death in fibroblast cells (Xiong and Parsons, 1997). Ethanol treatment increased PYK2 activity. Activation of PYK2 by stress signals has been reported to couple with JNK signaling pathway. It has been demonstrated that the tyrosine kinase PYK-2 is a critical intracellular signaling molecule, integrating chemokine and growth factor receptor stimulation with a variety of downstream pathways, including Ras, mitogen-activated protein (MAP) kinase, protein kinase C, and inositol phosphate metabolism (Lev et al., 1999). Chronic ethanol-induced increased ERK, JNK, and p38 activities might coordinate with changes in PYK2 and Akt-kinase activities. Akt, also known as protein kinase B (PKB) or RAC-PK (a protein kinase related to protein kinases A and C), is a serine-threonine kinase involved in signaling cascade in the survival of many cell types and is emerging as a key player in regulating anti-apoptotic events. Akt-protein kinase is the downstream target of phosphotidyl-3-inositol kinase (PI_3K) activation and the precise way in which 3-phosphoinositide generation induces Akt activation is still not completely understood. Decreased Akt activity after chronic ethanol treatment might enhance JNK activity and apoptosis in cerebellar granule cells (Shimoke et al., 1999). Another consequence of decreased Akt activity might be the activation of caspase-9 triggering the cell death pathway by ethanol. These studies imply that ethanol-induced changes in these kinases and perturbation of cell signaling pathways may contribute to impaired CNS development in fetal alcohol syndrome (FAS) and brain atrophy associated with alcoholic neurodegeneration. Further work needs to be directed towards identifying the genes positively or negatively influenced by chronic ethanol, as these may serve as putative therapeutic targets in modulating neuronal apoptosis in response to chronic ethanol neurotoxicity.

10.4 ALCOHOL NEUROTOXICOLOGY: NEW PERSPECTIVES FROM NEURAL STEM CELLS

Maternal ethanol consumption results in massive death of unique stage-specific cell populations of the developing brain, face, and cranial nerve ganglia, leading to an unrecoverable loss of tissue mass in these structures that might be responsible for FAS-associated malformations. As alcohol alters embryonic neurogenesis and fetal development, ethanol might affect neural stem cells (NSC) that are destined to differentiate into different neural cell types. Accordingly, some recent studies strengthen that acute ethanol consumption in a 4-day drinking model has profound effects on the proliferation of NSCs in adult brain (Crews et al., 2003). NSCs can give rise to different populations of neuronal cell types such as neurons, astrocytes, and oligodendrocytes by lineage restriction development (Figure 10.3). It is known that NSCs are derived from the cells lining the ventricles in brain. The lining of ventricles in the brain is divided into subventricular zone (SVZ) and ventricular zone (VZ). The VZ primarily houses the differentiating NSCs (Figure 10.4). Ethanol in particular has been reported to affect the proliferative NSCs in VZ (Miller, 1993; Miller, 1996). These reductions in VZ naturally contribute to the neuronal loss in the mature cortex. Limited ethanol exposure during embryogenesis selectively altered the proliferation of stem cells only during neurogenesis (Miller, 1996). In a further detailed study, Ma et al. (2003) have shown that ethanol disrupts NSC proliferation through inhibition of muscarinic signaling pathways. In studies using NSCs exposed to agonist and antagonists of muscarinic receptors, the results showed that ethanol blocks proliferation of NSCs that are stimulated by the activation of mAChR. It is also known that corticosteroids decrease NSC proliferation and alcohol is a known stress agent that increases corticosteroids (Jerrells et al., 1990). Similar to embryonic NSCs, adult NSCs also are affected by ethanol. Ethanol-induced inhibition of NSC proliferation and neurogenesis might underlie neurodegeneration and cognitive dysfunction in alcoholics.

Ongoing *in vitro* studies from our laboratory indicate that ethanol affects both proliferation and differentiation of neural stem cells isolated from the embryonic ventral mesencephalon area that eventually develop into mature substantia nigra, an area enriched with predominantly dopamine neurons that are implicated in alcoholic effects. Cells isolated from fetal brain regions can be grown into free-floating aggregate cultures as neurospheres. Neurospheres generated from different regions of embryonic brain can be expanded *ex vivo* over extended periods of time in culture (Hulspas and Quesenberry, 2000). Neurospheres contain a mix of multipotent stem cells as well as lineage restricted progenitor populations. In appropriate culture conditions and in the presence of growth factors, neurospheres retain the ability to

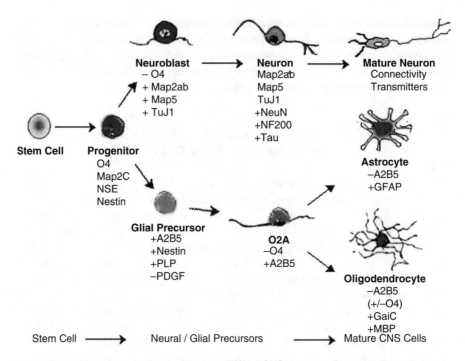

Figure 10.3 Development of stem cells into different CNS mature cell types by lineage restriction. The stage of neural and glial cell differentiation is associated with specific markers as indicated below each of the precursor cell types. Figure adapted from Holland, E.C. (2001), *Curr. Opin. Neurol.* 14, 683–688.

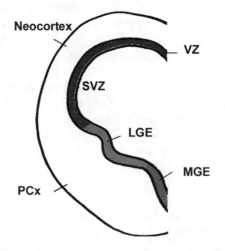

Figure 10.4 Hot spots of embryonic stem cell locations in embryonic brain development (E15.5). Stem cells are particularly enriched in the subventricular zone (SVZ) and the ventricular zone (VZ). Most of the stem cells located in lateral ganglionic eminence (LGE) and median ganglionic eminence (MGE) migrate towards striatum (Str) and cortex (PCx). Ethanol is known to affect the stem cells in both SVZ and VZ during development. Figure modified from Marin et al. (2001), *Nat. Rev. Neurosci.* 2, 780–790.

differentiate into neurons and glial cell types (Rao, 1999). Neuronal precursor cells (NPC) in neurospheres respond to environmental signals and accordingly NPC fates are altered. If the signals include an agent like ethanol, it is likely that it could alter NPC proliferation. This in turn impairs the normal developmental fate of NPCs, leading to improper neurogenesis and connectivity that contributes to memory deficits. Neurosphere cultures serve as a good model to assess the teratogenic or neurotoxicological effects of compounds such as ethanol on the survival, proliferation, and fate of cells derived from the neurospheres. Our study showed that ethanol affects proliferation and differentiation abilities of NPCs. Proliferation of NSCs is affected in a dose-dependent manner, while the differentiation of NSCs, particularly into astrocytes, is markedly reduced relative to differentiation into neurons, although differentiation into neuronal types is also affected (Figure 10.5). This reflects a basic

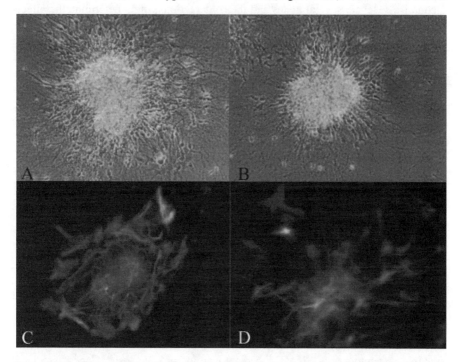

Figure 10.5 Effect of ethanol (50 m*M*) for 24 hours on the differentiation potential of neurospheres generated from rat ventral mesencephalon (E15.5). Neurospheres were generated as described in the text. Phase contrast views of neurospheres subjected to regular control differentiation medium (A) and medium plus ethanol 50 mM for 24 hours (B). Note the decline in overall neurite growth in the sphere. These spheres were fixed in 4% paraformaldehyde and triple labeled with fluorescence dyes for specific markers. Control neurospheres (C) and ethanol treated sphere (D). Neurons are stained red in color with MAP2 antibody conjugated to rhodopsin. Astrocytes are stained green in color with anti-glial fibrillary acidic protein attached to fluorochrome with FITC. Nuclei are stained blue with Hoechst 33342 as a counter stain. Note the compact healthy growth of the sphere in panel C versus diffusive and loose growth of cell types and reduced number of astrocytes due to ethanol treatment in panel D. Loss of astrocytes as well as neurons is due to ethanol-induced apoptosis, besides decreased proliferation and differentiation. (Shown in Color Figure 10.5 following page 212.)

faulty operation of stem cell turnover and altered cell fates of NSCs that might partly explain the shrinkage of the brain, the reduction in white matter, and altered embry-onic development in FAS. It might be possible in the future to intervene in FAS condition by *in utero* stem cell transplantation to overcome the cellular crisis in brain during fetal development. Stem cells would be powerful tools for regenerative and supplemental therapy, if used soon after the damage occurs in FAS.

10.5 GENES AND ALCOHOLISM: MICROARRAYS AND THE "ALCOHOL CHIP"

It has now become increasingly evident that there is a genetic component to alcoholism. Hence, several genes and their interactions might determine the effects of alcohol. Behavioral and biochemical responses mediating ethanol's actions have been difficult to study in humans and animals because of their complex polygenic nature, and attempts to isolate alcoholism genes have met with only modest success. However, in recent years, it has been possible to screen the entire genome for genes that mediate any given response to alcohol, an approach that has rapidly increased the number of candidate genes provisionally implicated in alcoholism. One exciting way to investigate alcohol responsive genes is through DNA microarrays using an "alcohol chip," identifying the genes that are either down-regulated or up-regulated. DNA microarrays are used to study the gene expression profiles by monitoring the hybridization of labelled nucleotide molecules to an array of complementary sequence of specific clusters of gene sequences immobilized to a substratum, usually on a silicon or glass slide. An alcohol chip consisting of several thousands of genes covering major cellular pathways such as signal transduction, apoptotic pathways, and xenobiotic metabolism was prepared, and the gene expression profiles were identified. The most striking changes are seen in genes coding for myelin protein, myelin-associated glycoprotein, genes involved in calcium, cAMP, and the thyroid signalling pathway (Lewohl et al., 2001; Mayfield et al., 2002). In a different study, $GABA_A$ receptor and ribosomal protein genes were also noticed to be highly altered after chronic ethanol treatment (Thibault et al., 2000).

Microarray analysis of cerebellar granule cell cultures exposed to ethanol showed significant gene expression changes in several vesicular docking proteins such as syntaxin, synaptobrenin, SNAP25, Synaptogamin, and Synapsin. Genes associated with vesicular transport such as dopamine transport (DAT), cytochrome P-4502E1, and two cyclin-dependent kinases were significantly down-regulated by ethanol (Crews, 2001). In a study that used high density microarray analysis, it was revealed that chronic ethanol treatment alters two classes of genes in the hippocampus. One group consisted of oxidoreductases, including ceruloplasmin, uricase, branched-chain alpha-keto acid dehydrogenase, NADH ubiquinone oxidoreductase, P450, NAD+-isocitrate dehydrogenase, and cytochrome *c* oxidase, which may be related to ethanol-induced oxidative stress. The other group of genes included ADP-ribosylation factor, RAS-related protein rab10, phosphatidylinositol 4-kinase, dynein-associated polypeptides, and dynamin-1 (Saito et al., 2002). A recent review has covered in more detail the relevance and utility of DNA microarrays in alcohol-related disease

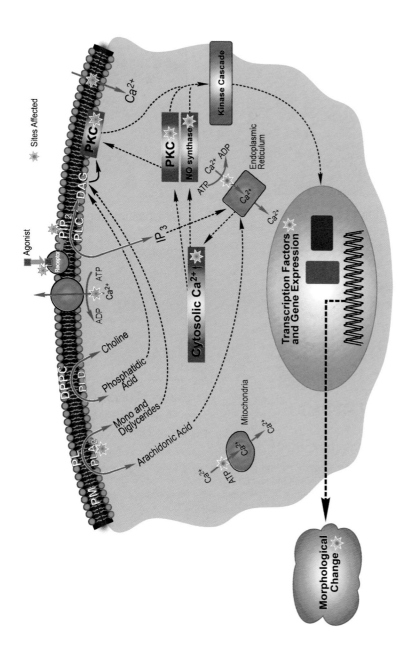

COLOR FIGURE 8.5 Schematic depicting changes in a number of intracellular signaling pathways and down-stream events leading to a structural change following exposure to a developmental neurotoxicant, polychlorinated biphenyls (PCBs). The stars represent the sites affected by PCBs *in vitro* and/or *in vivo*. (Adapted from Kodavanti and Tilson, 1997.)

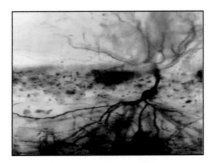

Hippocampal CA1 Pyramid

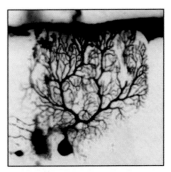

Cerebellum-Purkinje Cell

CA1 Spines

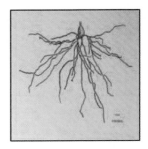

**Camera Lucida Drawing
of CA1 Basilar Tree**

COLOR FIGURE 8.6 Photomicrographs showing the morphology of neurons in hippocampus and cerebellum. Cerebral hemispheres were formalin-fixed for Rapid Golgi staining of tissue blocks which included the hippocampus and cerebellum, stained using the Golgi-Cox method. For branching and spine analysis of hippocampus, camera lucida drawings of the basilar dendritic tree were analyzed using the Sholl method of concentric circles. For dendritic spine analysis, counts were made along internal and terminal tip segments of neurons. For branching and spine analysis of cerebellar Purkinje cells, camera lucida drawings outlining the extent of the dendritic tree were made and the area of the dendritic domain was measured using a digitizing tablet. Dendritic branch density was assessed using an ocular grid. Dendritic spines were counted on tertiary terminal tip segments. (Courtesy of Dr. Ronald F. Mervis, Neurostructural Research Laboratory, Columbus, OH.)

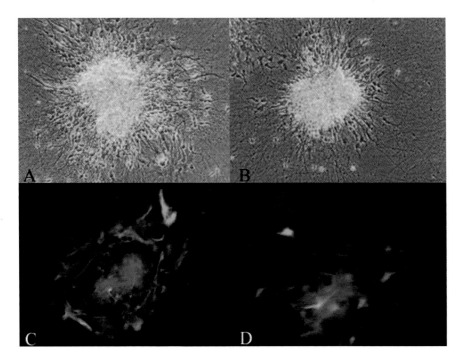

COLOR FIGURE 10.5 Effect of ethanol (50 m*M*) for 24 hours on the differentiation potential of neurospheres generated from rat ventral mesencephalon (E15.5). Neurospheres were generated as described in the text. Phase contrast views of neurospheres subjected to regular control differentiation medium (A) and medium plus ethanol 50 m*M* for 24 hours (B). Note the decline in overall neurite growth in the sphere. These spheres were fixed in 4% paraformaldehyde and triple labeled with fluorescence dyes for specific markers. Control neurospheres (C) and ethanol treated sphere (D). Neurons are stained red in color with MAP2 antibody conjugated to rhodopsin. Astrocytes are stained green in color with anti-glial fibrillary acidic protein attached to fluorochrome with FITC. Nuclei are stained blue with Hoechst 33342 as a counter stain. Note the compact healthy growth of the sphere in panel C versus diffusive and loose growth of cell types and reduced number of astrocytes due to ethanol treatment in panel D. Loss of astrocytes as well as neurons is due to ethanol-induced apoptosis besides decreased proliferation and differentiation.

in humans and animals (Crews, 2001). Microarray studies have just started in terms of alcoholism, and the results show important implications in understanding the molecular pathways involved in alcohol-related disease. Further studies in this direction hold great promise in unravelling the global gene expression in alcohol abuse and addiction.

10.6 TRANSGENIC AND GENE KNOCKOUT STUDIES

Recent progress in the creation of new animal models using recombinant DNA technology has provided a set of genetic tools by which the role of specific candidate genes in ethanol's actions can be examined. These techniques include the creation of transgenic and null mutant mice, and researchers attempting to elucidate the roles of specific genes in alcoholism risk have benefited from advances in genetic engineering and utilized genetically modified animals to understand the behavioral and pathological effects of ethanol. Two important tools used by researchers include transgenic mice, in which a foreign gene is integrated into an animal's genetic material, and knockout/knockin mice, in which targeted genes either are rendered nonfunctional or are altered. Both of these animal models are currently used in alcohol research to determine how genes may influence the development of alcoholism in humans. The use of transgenic animals that overexpress particular genes, or of null mutant (knockout) animals, in which particular genes have been eliminated, provides an unprecedented opportunity to elucidate the targets of ethanol's actions.

Several genes have recently been targeted for their potential influence on alcohol responses (Browman and Crabbe, 1999; Hoffman et al., 2001). Many *in vivo* and *in vitro* data implicate the GABA-A receptor in the actions of alcohol, although the detailed interactions of alcohol with this receptor complex are not clear. Nevertheless, studies of knockout mice have increased our knowledge of the action of this receptor complex with respect to alcohol, and future efforts to identify specific proteins that mediate subunit composition will help to identify the mechanism of action of alcohol. Furthermore, several studies have demonstrated that the effects of alcohol on ionotropic and metabotropic membrane receptors depend on protein–phosphorylation pathways. Alcohol might exert an indirect effect on phosphatases to change receptor function by altering phosphorylation sites. Alternatively, phosphorylation might mediate the sensitivity of receptors to the direct effects of alcohol. It is not clear which of these two hypotheses, if either, is correct. In a manipulation of the glutamatergic system, mice that lack the gene encoding *fyn*, a nonreceptor tyrosine kinase, had increased sensitivity to the sedative effects of alcohol. In the absence of *fyn*, the normal alcohol-stimulated phosphorylation of NMDA receptors is absent. Hippocampal NMDA receptors from *fyn* null mutant mice fail to develop acute tolerance to alcohol *in vitro*. Although a specific role for *fyn* in modulating the response to alcohol is not yet clear, the finding that acute and chronic alcohol administration affects NMDA receptors suggests that this is a good candidate for further investigation. As knockouts are developed that target protein kinases or other signal-transduction proteins, they can be used to further investigate these questions. For example, null mutants that lack the protein kinase C gamma

Table 10.1   Genetically Altered Mice Used in Alcohol Research

Genes Targeted	Phenotypic Effects	References
GABA A Receptor Subunits		
Gamma2L subunit	No difference in alcohol-potentiated response to GABA in neuronal tissue. No differences in sensitivity to hypnotic effects of alcohol, tolerance or dependence.	(Homanics et al., 1998)
Alpha6 subunit	No difference in hypnotic sensitivity to alcohol. No difference in acute functional tolerance, protracted tolerance or withdrawal hyper-excitability.	(Homanics et al., 1997)
Beta3 subunit	No difference in sensitivity to hypnotic effects of alcohol.	(Homanics et al., 1997; Homanics et al., 1998)
Dopamine Receptor Subunits		
Dopamine D2 receptor	Decreased alcohol consumption, insensitivity to locomotor-depressant receptor effect of alcohol, reduced sensitivity to the ataxic effects of alcohol.	(Phillips et al., 1998)
Dopamine D4 receptor	Increased sensitivity to locomotor stimulant effects of alcohol.	(Rubinstein et al., 1997)
Other Genes		
5-Hydroxytryptamine HT1B receptor	Increased alcohol consumption, decreased sensitivity to alcohol-induced ataxia.	(Crabbe et al., 1996)
PKC gamma2 subunit	Less sensitive to hypnotic and hypothermic effects of alcohol.	(Harris et al., 1995)
Fyn tyrosine kinase	Increased sensitivity to alcohol-induced loss-of-righting reflex.	(Pierce and Kalivas, 1997)
Neuropeptide Y (null mutant)	Increased alcohol consumption.	(Thiele et al., 2000)
NeuropeptideY (transgenic overexpression)	Decreased alcohol consumption.	
TGF-alpha (transgenic overexpression)	Increased sensitivity to sedative effects of alcohol.	(Hilakivi-Clarke et al., 1993)

(PKC-γ) isoform show decreased ethanol-induced hypothermia and loss of the righting reflex compared to control animals. These mice should be useful in investigating the role of this kinase in mediating ethanol sensitivity. The mesolimbic dopamine system might be important in mediating the effects of many addictive drugs. Among the projections of the mesolimbic system, those to the nucleus accumbens could be particularly important, and information from several regions converges on the nucleus accumbens. Although dopamine is the predominant neurotransmitter of the nucleus accumbens, other neurotransmitter systems, such as the serotonin and GABA systems, are in a position to modulate function in this circuit. Several knockout mice for dopamine-specific receptor subtypes are available now and have been utilized in alcohol research. Several genes have very recently been targeted for their potential influence on alcohol responses, which is summarized in Table 10.1.

The use of transgenic mice to explore the mechanisms of alcohol action is just beginning. One of the first studies in this direction by Hilakivi-Clarke identified the effect of alcohol on behavior in a highly aggressive transgenic TGF alpha male mouse (Hilakivi-Clarke et al., 1993). These results suggested that male TGF alpha mice have physiological sensitivity to alcohol. Then, transgenic mice overexpressing the neuropeptide Y (NPY) gene were developed to study the alcohol consumption pattern since alcohol-preferring rats have lower levels of NPY in several brain regions (Thiele et al., 1998). In parallel, they also included NPY-deficient mice for a parallel comparison and reported increased consumption in NPY-deficient mice. NPY-deficient mice were also less sensitive to the sedative/hypnotic effects of ethanol. In contrast, transgenic mice that overexpressed a marked NPY gene in neurons that usually express it had a lower preference for ethanol and were more sensitive to the sedative/hypnotic effects of this drug than controls. These findings suggested direct evidence that alcohol consumption and resistance are inversely related to NPY levels in the brain. Subsequently, Heaton's group demonstrated that overexpression of the anti-apoptotic gene Bcl-2 protects the cerebellum from ethanol neurotoxicity in neonatal mice (Heaton et al., 1999). This study suggested that modulation of cell death effect or repressor gene products may play a significant role in developmental ethanol neurotoxicity. The same group further reported that overexpression of nerve growth factor gene in mice ameliorated ethanol neurotoxicity in the developing cerebellum (Heaton et al., 2000). These results suggested that endogenous over-expression of neurotrophic factors, which have previously been shown to protect against ethanol neurotoxicity in culture, can serve a similar protective function in the intact animal.

Genetically altered animals have led to exciting advances in our understanding of how alcohol exerts its effects *in vivo* at the whole animal level. It is clear that many neuronal signaling pathways are involved. Transgenic technologies coupled with more traditional types of analysis (e.g., molecular, electrophysiological) result in rapid progress in unraveling the complexities of alcohol action. There are many knockout mice that have not yet been tested for their response to alcohol, but the exciting advances obtained from work to date demonstrate the importance of this technique. Other genes might compensate for the loss of any one gene during development. Thus, it might be difficult to interpret the effects of eliminating a single gene, but recent advances are promising. Furthermore, the development of inducible knockouts, in which gene deletion can occur at any point during the lifetime of an animal, should lead to rapid advances in elucidating the genetics of alcoholism. There is no doubt that this type of approach in combination with candidate gene searching and targeted search for specific gene expression will provide a new avenue for revealing the mechanism that underlies various alcohol-related disorders and for developing new therapeutic strategies. Scientific data from human brain materials, based on data from experimental animal models, will be warranted. Although there are caveats to these approaches, alcohol researchers have been in the forefront of this exciting area of research and have made important discoveries. However, one outstanding question that remains is whether these new transgenic technologies (i.e., conditional knockouts) would allow us to detect a wider range of candidate genes, telling us more about the neurobiological mediators of responses to alcohol. Therefore, it is quite likely in the near future that the investigators in alcohol research would utilize these powerful genetic manipulation technologies to investigate the role of specific candidate genes possibly involved in predisposition to alcoholism. An understanding of these target genes not only provides insight into genetically mediated differences in susceptibility to ethanol's effects, but may also lead to the development of therapies for alcohol abuse and alcoholism and interventions in the pathological effects of ethanol.

10.7 CONCLUDING REMARKS AND FUTURE DIRECTIONS

Although remarkable advances have been made in the neurotoxicology of alcohol action in the past decade, a number of unsolved fundamental questions still remain. The mechanistic action of alcohol leading to addiction is a major issue that must be clarified, not only because of scientific interest but also to increase the means of effective treatment. As more genes are identified in animals and humans, the outstanding issue that needs to be addressed is the available options for treatment and prospects for pharmacogenomics in alcohol research. Further, it would be necessary to delineate which of the genes identified so far in animals are relevant for alcohol-related phenotypes in humans. Furthermore, it will be particularly important to determine whether the new transgenic technologies (i.e., antisense, conditional knockouts, and RNA interference) would allow us to detect a wider range of candidate genes, telling us more about the neurobiological mediators of responses to alcohol. Also, can these genetic studies help us to unravel the complex commonalities of alcohol dependence, antisocial personality, and depression? Further accumulation of new knowledge regarding the type of cell death, the specific neurotransmitter receptors involved, and their downstream consequences leading to a specific signaling cascade, and utilization of knowledge from new disciplines such as stem cell biology and microarrays will continue to provide better opportunities to address alcohol toxicity effectively. New studies are using combinatorial chemistry, high throughput screening, and genetic and gene delivery approaches to better understand the development of treatment paradigms in alcoholism. Finally, there is every good reason to be optimistic that these types of investigations will lead to novel and efficacious therapies for alcohol-related diseases.

REFERENCES

Arendt, T. et al. (1988), Loss of neurons in the rat basal forebrain cholinergic projection system after prolonged intake of ethanol, *Brain Res. Bull.* 21, 563–569.

Arendt, T. et al. (1995), Degeneration of rat cholinergic basal forebrain neurons and reactive changes in nerve growth factor expression after chronic neurotoxic injury — II. Reactive expression of the nerve growth factor gene in astrocytes, *Neuroscience* 65, 647–659.

Bannigan, J., and Burke, P. (1982), Ethanol teratogenicity in mice: a light microscopic study, *Teratology* 26, 247–254.

Barrett, G. L., and Georgiou, A. (1996), The low-affinity nerve growth factor receptor p75NGFR mediates death of PC12 cells after nerve growth factor withdrawal, *J. Neurosci. Res.* 45, 117–128.

Bhave, S. V., and Hoffman, P. L. (1997), Ethanol promotes apoptosis in cerebellar granule cells by inhibiting the trophic effect of NMDA, *J. Neurochem.* 68, 578–586.

Blitzer, R. D., Gil, O., and Landau, E. M. (1990), Long-term potentiation in rat hippocampus is inhibited by low concentrations of ethanol, *Brain Res.* 537, 203–208.

Boulton, T. G. et al. (1991), ERKs: a family of protein-serine/threonine kinases that are activated and tyrosine phosphorylated in response to insulin and NGF, *Cell* 65, 663–675.

Brennan, C. H., Crabbe, J., and Littleton, J. M. (1990), Genetic regulation of dihydropyridine-sensitive calcium channels in brain may determine susceptibility to physical dependence on alcohol, *Neuropharmacology* 29, 429–432.

Brooks, P. J. (2000), Brain atrophy and neuronal loss in alcoholism: a role for DNA damage? *Neurochem. Int.* 37, 403–412.

Browman, K. E., and Crabbe, J. C. (1999), Alcohol and genetics: new animal models, *Mol. Med. Today* 5, 310–318.

Castoldi, A. F. et al. (1998), Ethanol selectively interferes with the trophic action of NMDA and carbachol on cultured cerebellar granule neurons undergoing apoptosis, *Brain Res. Dev. Brain Res.* 111, 279–289.

Charness, M. E. (1993), Brain lesions in alcoholics, *Alcohol Clin. Exp. Res.* 17, 2–11.

Charness, M. E., Querimit, L. A., and Henteleff, M. (1988), Ethanol differentially regulates G proteins in neural cells, *Biochem. Biophys. Res. Commun.* 155, 138–143.

Climent, E. et al. (2002), Ethanol exposure enhances cell death in the developing cerebral cortex: role of brain-derived neurotrophic factor and its signaling pathways, *J. Neurosci. Res.* 68, 213–225.

Cohen-Kerem, R., and Koren, G. (2003), Antioxidants and fetal protection against ethanol teratogenicity. I. Review of the experimental data and implications to humans, *Neurotoxicol. Teratol.* 25, 1–9.

Crabbe, J. C. et al. (1996), Elevated alcohol consumption in null mutant mice lacking 5-HT1B serotonin receptors, *Nat. Genet.* 14, 98–101.

Crews, F. T. (2001), Summary report of a symposium: genes and gene delivery for diseases of alcoholism, *Alcohol Clin. Exp. Res.* 25, 1778–1800.

Crews, F. T. et al. (1998), Ethanol, stroke, brain damage, and excitotoxicity, *Pharmacol. Biochem. Behav.* 59, 981–991.

Crews, F. T. et al. (2003), Neural stem cells and alcohol, *Alcohol. Clin. Exp. Res.* 27, 324–335.

Daniell, L. C., Brass, E. P., and Harris, R. A. (1987), Effect of ethanol on intracellular ionized calcium concentrations in synaptosomes and hepatocytes, *Mol. Pharmacol.* 32, 831–837.

Diamond, I., and Gordon, A. S. (1997), Cellular and molecular neuroscience of alcoholism, *Physiol. Rev.* 77, 1–20.

Dikic, I. et al. (1996), A role for Pyk2 and Src in linking G-protein-coupled receptors with MAP kinase activation, *Nature* 383, 547–550.

Dildy, J. E., and Leslie, S. W. (1989), Ethanol inhibits NMDA-induced increases in free intracellular Ca2+ in dissociated brain cells, *Brain Res.* 499, 383–387.

Ellis, R. E., Yuan, J. Y., and Horvitz, H. R. (1991), Mechanisms and functions of cell death, *Annu. Rev. Cell Biol.* 7, 663–698.

Fagan, A. M. et al. (1996), TrkA, but not TrkC, receptors are essential for survival of sympathetic neurons in vivo, *J. Neurosci.* 16, 6208–6218.

Fink, K., and Gothert, M. (1990), Inhibition of N-methyl-D-aspartate-induced noradrenaline release by alcohols is related to their hydrophobicity, *Eur. J. Pharmacol.* 191, 225–229.

Freund, G. (1994), Apoptosis and gene expression: perspectives on alcohol-induced brain damage, *Alcohol* 11, 385–387.

Goldstein, D. B., and Chin, J. H. (1981), Interaction of ethanol with biological membranes, *Fed. Proc.* 40, 2073–2076.

Gordon, A. S., Collier, K., and Diamond, I. (1986), Ethanol regulation of adenosine receptor-stimulated cAMP levels in a clonal neural cell line: an *in vitro* model of cellular tolerance to ethanol, *Proc. Natl. Acad. Sci. USA* 83, 2105–2108.

Hamby-Mason, R. et al. (1997), Catalase mediates acetaldehyde formation from ethanol in fetal and neonatal rat brain, *Alcohol. Clin. Exp. Res.* 21, 1063–1072.

Hamre, K. M., and West, J. R. (1993), The effects of the timing of ethanol exposure during the brain growth spurt on the number of cerebellar Purkinje and granule cell nuclear profiles, *Alcohol. Clin. Exp. Res.* 17, 610–622.

Harper, C. (1998), The neuropathology of alcohol-specific brain damage, or does alcohol damage the brain? *J. Neuropathol. Exp. Neurol.* 57, 101–110.

Harper, C., and Kril, J. (1991), If you drink your brain will shrink. Neuropathological considerations, *Alcohol Alcohol Suppl.* 1, 375–380.

Harris, R. A., and Allan, A. M. (1989), Alcohol intoxication: ion channels and genetics, *FASEB J.* 3, 1689–1695.

Harris, R. A. et al. (1995), Mutant mice lacking the gamma isoform of protein kinase C show decreased behavioral actions of ethanol and altered function of gamma-aminobutyrate type A receptors, *Proc. Natl. Acad. Sci. USA* 92, 3658–3662.

Haviryaji, K. S., and Vemuri, M. C. (1997), Effect of ethanol on nuclear casein kinase II activity in brain, *Neurochem. Res.* 22, 699–704.

Heaton, M. B. et al. (1999), Bcl-2 overexpression protects the neonatal cerebellum from ethanol neurotoxicity, *Brain Res.* 817, 13–18.

Heaton, M. B., Mitchell, J. J., and Paiva, M. (2000), Overexpression of NGF ameliorates ethanol neurotoxicity in the developing cerebellum, *J. Neurobiol.* 45, 95–104.

Hilakivi-Clarke, L., Durcan, M., and Goldberg, R. (1993), Effect of alcohol on elevated aggressive behavior in male transgenic TGF alpha mice, *Neuroreport* 4, 155–158.

Hillbom, M. (1999), Oxidants, antioxidants, alcohol and stroke, *Front. Biosci.* 4, e67–71.

Hinson, W. G. et al. (1991), Nuclear protein phosphorylation in rat cerebral cells following acute exposure to ethanol, *Appl. Theor. Electrophor.* 2, 93–99.

Hoffman, P. L. et al. (2001), Transgenic and gene "knockout" models in alcohol research, *Alcohol Clin. Exp. Res.* 25, 60S–66S.

Holland, E.C. (2001), Progenitor cells and glioma formation, *Curr. Opin. Neurol.* 14, 683–688.

Homanics, G. E. et al. (1997), Gene knockout of the alpha6 subunit of the gamma-aminobutyric acid type A receptor: lack of effect on responses to ethanol, pentobarbital, and general anesthetics, *Mol. Pharmacol.* 51, 588–596.

Homanics, G. E. et al. (1998), Alcohol and anesthetic mechanisms in genetically engineered mice, *Front. Biosci.* 3, D548–558.

Hulspas, R., and Quesenberry, P. J. (2000), Characterization of neurosphere cell phenotypes by flow cytometry, *Cytometry* 40, 245–250.

Ikonomidou, C. et al. (2000), Ethanol-induced apoptotic neurodegeneration and fetal alcohol syndrome, *Science* 287, 1056–1060.

Iorio, K. R. et al. (1992), Chronic exposure of cerebellar granule cells to ethanol results in increased N-methyl-D-aspartate receptor function, *Mol. Pharmacol.* 41, 1142–1148.

Jensen, G. B., and Pakkenberg, B. (1993), Do alcoholics drink their neurons away? *Lancet* 342, 1201–1204.

Jerrells, T. R., Smith, W., and Eckardt, M. J. (1990), Murine model of ethanol-induced immunosuppression, *Alcohol Clin. Exp. Res.* 14, 546–550.

Kril, J. J., and Halliday, G. M. (1999), Brain shrinkage in alcoholics: a decade on and what have we learned? *Prog. Neurobiol.* 58, 381–387.

Kril, J. J. et al. (1997), The cerebral cortex is damaged in chronic alcoholics, *Neuroscience* 79, 983–998.

Kroemer, G. et al. (1995), The biochemistry of programmed cell death, *FASEB J.* 9, 1277–1287.

Lev, S. et al. (1995), Protein tyrosine kinase PYK2 involved in Ca(2+)-induced regulation of ion channel and MAP kinase functions, *Nature* 376, 737–745.

Lev, S. et al. (1999), Identification of a novel family of targets of PYK2 related to Drosophila retinal degeneration B (rdgB) protein, *Mol. Cell Biol.* 19, 2278–2288.

Lewohl, J. M. et al. (2001), Application of DNA microarrays to study human alcoholism, *J. Biomed. Sci.* 8, 28–36.

Li, Z., Miller, M. W., and Luo, J. (2002), Effects of prenatal exposure to ethanol on the cyclin-dependent kinase system in the developing rat cerebellum, *Brain Res. Dev. Brain Res.* 139, 237–245.

Luo, J., and Miller, M. W. (1997), Differential sensitivity of human neuroblastoma cell lines to ethanol: correlations with their proliferative responses to mitogenic growth factors and expression of growth factor receptors, *Alcohol Clin. Exp. Res.* 21, 1186–1194.

Luo, J., West, J. R., and Pantazis, N. J. (1997), Nerve growth factor and basic fibroblast growth factor protect rat cerebellar granule cells in culture against ethanol-induced cell death, *Alcohol Clin. Exp. Res.* 21, 1108–1120.

Ma, W. et al. (2003), Ethanol blocks both basic fibroblast growth factor- and carbachol-mediated neuroepithelial cell expansion with differential effects on carbachol-activated signaling pathways, *Neuroscience* 118, 37–47.

Machu, T., Woodward, J. J., and Leslie, S. W. (1989), Ethanol and inositol 1,4,5-trisphosphate mobilize calcium from rat brain microsomes, *Alcohol* 6, 431–436.

MacLennan, A. J., Lee, N., and Walker, D. W. (1995), Chronic ethanol administration decreases brain-derived neurotrophic factor gene expression in the rat hippocampus, *Neurosci. Lett.* 197, 105–108.

Mahadev, K., and Vemuri, M. C. (1998a), Effect of ethanol on chromatin and nonhistone nuclear proteins in rat brain, *Neurochem. Res.* 23, 1179–1184.

Mahadev, K., and Vemuri, M. C. (1998b), Selective changes in protein kinase C isoforms and phosphorylation of endogenous substrate proteins in rat cerebral cortex during pre- and postnatal ethanol exposure, *Arch. Biochem. Biophys.* 356, 249–257.

Mantle, D., and Preedy, V. R. (1999), Free radicals as mediators of alcohol toxicity, *Adverse Drug React. Toxicol. Rev.* 18, 235–252.

Marcussen, B. L. et al. (1994), Developing rat Purkinje cells are more vulnerable to alcohol-induced depletion during differentiation than during neurogenesis, *Alcohol* 11, 147–156.

Marin, O., Rubenstein, J.L., and Long, A. (2001), Remarkable journey: tangential migration in the telencephalon, *Nat. Rev. Neurosci.* 2, 780–790.

Marks, N., and Berg, M. J. (1999), Recent advances on neuronal caspases in development and neurodegeneration, *Neurochem. Int.* 35, 195–220.

Mayfield, R. D. et al. (2002), Patterns of gene expression are altered in the frontal and motor cortices of human alcoholics, *J. Neurochem.* 81, 802–813.

Messing, R. O. et al. (1986), Ethanol regulates calcium channels in clonal neural cells, *Proc. Natl. Acad. Sci. USA* 83, 6213–6215.

Messing, R. O., Sneade, A. B., and Savidge, B. (1990), Protein kinase C participates in upregulation of dihydropyridine-sensitive calcium channels by ethanol, *J. Neurochem.* 55, 1383–1389.

Michaelis, E. K. (1990), Fetal alcohol exposure: cellular toxicity and molecular events involved in toxicity, *Alcohol Clin. Exp. Res.* 14, 819–826.

Miller, M. W. (1993), Migration of cortical neurons is altered by gestational exposure to ethanol, *Alcohol Clin. Exp. Res.* 17, 304–314.

Miller, M. W. (1996), Limited ethanol exposure selectively alters the proliferation of precursor cells in the cerebral cortex, *Alcohol Clin. Exp. Res.* 20, 139–143.

Mitchell, J. J., Paiva, M., and Heaton, M. B. (1999a), The antioxidants vitamin E and beta-carotene protect against ethanol-induced neurotoxicity in embryonic rat hippocampal cultures, *Alcohol* 17, 163–168.

Mitchell, J. J., Paiva, M., and Heaton, M. B. (1999b), Vitamin E and beta-carotene protect against ethanol combined with ischemia in an embryonic rat hippocampal culture model of fetal alcohol syndrome, *Neurosci. Lett.* 263, 189–192.

Morrow, A. L. et al. (1990), Chronic ethanol and pentobarbital administration in the rat: effects on GABAA receptor function and expression in brain, *Alcohol* 7, 237–244.

Namura, S. et al. (1998), Activation and cleavage of caspase-3 in apoptosis induced by experimental cerebral ischemia, *J. Neurosci.* 18, 3659–3668.

Napper, R. M., and West, J. R. (1995), Permanent neuronal cell loss in the cerebellum of rats exposed to continuous low blood alcohol levels during the brain growth spurt: a stereological investigation, *J. Comp. Neurol.* 362, 283–292.

Nuydens, R. et al. (1998), Okadaic acid-induced apoptosis in neuronal cells: evidence for an abortive mitotic attempt, *J. Neurochem.* 70, 1124–1133.

Oberdoerster, J., Kamer, A. R., and Rabin, R. A. (1998), Differential effect of ethanol on PC12 cell death, *J. Pharmacol. Exp. Ther.* 287, 359–365.

Olney, J. W. et al. (2002a), Ethanol-induced caspase-3 activation in the *in vivo* developing mouse brain, *Neurobiol. Dis.* 9, 205–219.

Olney, J. W. et al. (2002b), The enigma of fetal alcohol neurotoxicity, *Ann. Med.* 34, 109–119.

Pellegrino, S. M., and Druse, M. J. (1992), The effects of chronic ethanol consumption on the mesolimbic and nigrostriatal dopamine systems, *Alcohol Clin. Exp. Res.* 16, 275–280.

Pfefferbaum, A. et al. (1997), Frontal lobe volume loss observed with magnetic resonance imaging in older chronic alcoholics, *Alcohol Clin. Exp. Res.* 21, 521–529.

Pfefferbaum, A. et al. (1998), A controlled study of cortical gray matter and ventricular changes in alcoholic men over a 5-year interval, *Arch. Gen. Psychiatry* 55, 905–912.

Phillips, T. J. et al. (1998), Alcohol preference and sensitivity are markedly reduced in mice lacking dopamine D2 receptors, *Nat. Neurosci.* 1, 610–615.

Pierce, R. C., and Kalivas, P. W. (1997), A circuitry model of the expression of behavioral sensitization to amphetamine-like psychostimulants, *Brain Res. Brain Res. Rev.* 25, 192–216.

Puntarulo, S., and Cederbaum, A. I. (1989), Interactions between paraquat and ferric complexes in the microsomal generation of oxygen radicals, *Biochem. Pharmacol.* 38, 2911–2918.

Rabin, R. A. (1993), Ethanol-induced desensitization of adenylate cyclase: role of the adenosine receptor and GTP-binding proteins, *J. Pharmacol. Exp. Ther.* 264, 977–983.

Rajgopal, Y., and Vemuri, M. C. (2001), Ethanol induced changes in cyclin-dependent kinase-5 activity and its activators, P35, P67 (Munc-18) in rat brain, *Neurosci. Lett.* 308, 173–176.

Rajgopal, Y., and Vemuri, M. C. (2002), Calpain activation and alpha-spectrin cleavage in rat brain by ethanol, *Neurosci. Lett.* 321, 187–191.

Rajgopal, Y., Chetty, C.S., and Vemuri, M.C. (2003), Differential modulation of apoptosis associated proteins by ethanol in rat cerebral cortex and cerebellum, *Eur. J. Pharmacol.* 470, 117–124.

Rao, M. S. (1999), Multipotent and restricted precursors in the central nervous system, *Anat. Rec.* 257, 137–148.

Renis, M. et al. (1996), Nuclear DNA strand breaks during ethanol-induced oxidative stress in rat brain, *FEBS Lett.* 390, 153–156.

Resnicoff, M. et al. (1994), Ethanol inhibits insulin-like growth factor-1-mediated signalling and proliferation of C6 rat glioblastoma cells, *Lab. Invest.* 71, 657–662.

Robaye, B. et al. (1994), Apoptotic cell death analyzed at the molecular level by two-dimensional gel electrophoresis, *Electrophoresis* 15, 503–510.

Rubinstein, M. et al. (1997), Mice lacking dopamine D4 receptors are supersensitive to ethanol, cocaine, and methamphetamine, *Cell* 90, 991–1001.

Saito, M. et al. (2002), Microarray analysis of gene expression in rat hippocampus after chronic ethanol treatment, *Neurochem. Res.* 27, 1221–1229.

Shimoke, K. et al. (1999), Inhibition of phosphatidylinositol 3-kinase activity elevates c-Jun N-terminal kinase activity in apoptosis of cultured cerebellar granule neurons, *Brain Res. Dev. Brain Res.* 112, 245–253.

Su, B., and Karin, M. (1996), Mitogen-activated protein kinase cascades and regulation of gene expression, *Curr. Opin. Immunol.* 8, 402–411.

Tabakoff, B., Munoz-Marcus, M., and Fields, J. Z. (1979), Chronic ethanol feeding produces an increase in muscarinic cholinergic receptors in mouse brain, *Life Sci.* 25, 2173–2180.

Tanaka, H. et al. (1988), Fetal alcohol syndrome in rats: conditions for improvement of ethanol effects on fetal cerebral development with supplementary agents, *Biol. Neonate* 54, 320–329.

Thibault, C. et al. (2000), Expression profiling of neural cells reveals specific patterns of ethanol-responsive gene expression, *Mol. Pharmacol.* 58, 1593–1600.

Thiele, T. E. et al. (1998), Ethanol consumption and resistance are inversely related to neuropeptide Y levels, *Nature* 396, 366–369.

Thiele, T. E. et al. (2000), Neurobiological responses to ethanol in mutant mice lacking neuropeptide Y or the Y5 receptor, *Pharmacol. Biochem. Behav.* 67, 683–691.

Victor, M. (1994), Alcoholic dementia, *Can. J. Neurol. Sci.* 21, 88–99.

Wegelius, K., and Korpi, E. R. (1995), Ethanol inhibits NMDA-induced toxicity and trophism in cultured cerebellar granule cells, *Acta Physiol. Scand.* 154, 25–34.

Woodward, J. J., and Gonzales, R. A. (1990), Ethanol inhibition of N-methyl-D-aspartate-stimulated endogenous dopamine release from rat striatal slices: reversal by glycine, *J. Neurochem.* 54, 712–715.

Xia, Z. et al. (1995), Opposing effects of ERK and JNK-p38 MAP kinases on apoptosis, *Science* 270, 1326–1331.

Xiong, W., and Parsons, J. T. (1997), Induction of apoptosis after expression of PYK2, a tyrosine kinase structurally related to focal adhesion kinase, *J. Cell Biol.* 139, 529–539.

Zhang, F. X., Rubin, R., and Rooney, T. A. (1998), Ethanol induces apoptosis in cerebellar granule neurons by inhibiting insulin-like growth factor 1 signaling, *J. Neurochem.* 71, 196–204.

Index

A

Aβ protein, 113
  in Alzheimer's disease, 108–109
Abnormal hind-brain development,
    association with alteration in
    Egr-2 expression, 56
Academic development, impaired by
    developmental lead exposure,
    185, 152
Acetylation, association with transcription
    activation, 19
Acquired hepatocerebral degeneration, 200
Activation, 17
  associated with acetylation, 19
  determined by nuclear run-on assays, 23
Acute phase proteins, regulated by NF-κB,
    67
Adenine, 13
Adhesion molecules, regulated by NF-κB,
    67
Aerobic glucose metabolism, required by
    CNS, 155
Alcohol. *See also* Ethanol
  as developmental neurotoxicant,
    162–163, 160
  effects of developmental exposure to,
    162–163
  mechanisms of neurotoxicity, 203–209
  neurotoxicity of, 201–203
  vulnerability of developing brain to, 152
Alcoholic dementia, 202–203
Alcoholism, genetic basis of, 212–213
Alcohol-related neuronal loss (ARNL), 202
Algicides, presence of alkyltin compounds
    in, 73

Alkyltin compounds, sources of, 73
Alpha synuclein, 109
Aluminum exposure
  enhancing cerebral pro-oxidant status,
    113
  exacerbating oxidative stress, 113
  and incidence of Alzheimer's disease,
    110
Alzheimer's disease, 134
  altered expression of NF-κB in, 72
  application of TMT syndrome research
    to, 76
  association with low folic acid levels,
    113
  Aβ protein in, 108–109
  dietary components of, 113
  early onset in Down's syndrome, 112
  genes altered in transgenic animal
    models of, 115
  impaired cytochrome c oxidase activity
    in, 135
  role of cytokines in development of,
    114–115
Ammunition, presence of lead in, 184
Amyloid deposition, 108
Amyloid precursor protein (APP), 108
Amyotrophic lateral sclerosis (ALS), 134,
    109
  cytochrome c oxidase mutations in, 135
  genes altered in transgenic animal
    models of, 115
  role of microglial activation in motor-
    associated neuron death, 114
Anorexia, association with TMT syndrome,
    73
Antiapoptosis genes, regulated by NF-κB,
    67

223